INSIDE THE BRAIN

RONALD KOTULAK

INSIDE THE BRAIN

Revolutionary Discoveries of How the Mind Works

Andrews and McMeel
A Universal Press Syndicate Company
Kansas City

For information write to
Andrews and McMeel, a Universal Press Syndicate Company,
4520 Main Street, Kansas City, Missouri 64111.

First Printing, May 1996
Third Printing, November 1996

Library of Congress Cataloging-in-Publication Data
Kotulak, Ronald, 1935–
Inside the brain : revolutionary discoveries of how the mind works /
Ronald Kotulak.
p. cm.
Includes index.
ISBN: 0-8362-1043-3 (hd)
1. Brain. 2. Neuropsychology. 3. Neurosciences. I. Title.
QP376.K666
612.8'2—dc20 95-45884
CIP

Attention: Schools and Businesses
Andrews and McMeel books are available at quantity discounts with bulk purchase for educational, business, or sales promotional use. For information, please write to Special Sales Department, Andrews and McMeel, Main Street, Kansas City, Missouri 64111.

To all parents and teachers
and especially to my wife, Donna,
and our children, Jeff, Kerry, Chris, Paul, and Lisa.

Contents

Acknowledgments

Such a complex undertaking as *Inside the Brain* could not have succeeded without the help of many people, including the hundreds of scientists I interviewed who gave generously of their time, insights, and sense of adventure. At the *Tribune* I would particularly like to thank Jack Fuller, for planting the seed of the project; Howard Tyner, for his strong support; Jim O'Shea, for his enthusiastic encouragement; Marshall Froker, for his skillful editing; and Terry Volpp and other members of the graphics team, for their imagination. I would also like to thank Donna Martin and Harriet Choice at Andrews and McMeel for their vision of the book and their grace in seeing it through.

Introduction

Like many major investigations, *Inside the Brain* started with a question: "Why do some children turn out bad?"

I was asked this question by the editor of the *Chicago Tribune*, who was disturbed about the increasing rate of violence among young people. The *Tribune* had undertaken a year-long series documenting the lives and deaths of children aged fourteen years and younger who were killed in the Chicago area in 1993, and the editor was frustrated with the findings.

Case after case of the sixty-one children who met violent deaths repeated the same disheartening story. The facts were chilling, but they did not lead to any better understanding of the problem or possible ways to prevent it. Usually the children were born to teenage mothers. There was no father at home. The children were abused and impoverished, and they lived in bad neighborhoods. Some were both victims and perpetrators of violence. Yet most children living under similar conditions do not turn out bad.

There had to be something more that could help explain why some children were turning to violence. Meeting me in the hall one day, the editor, Jack Fuller, who is now publisher, asked if there was anything going on in brain research that could provide some answers.

A decade ago I would have said no. The brain was considered a "black box" then and there was little hope of ever figuring out how it worked. Now, however, we can ask: "How does the brain work?" We can ask it because we are riding the crest of a revolution in molecular biology and genetics that is opening the door to understanding the brain. Along with new tools of imaging technology, which can "see" the chemical traces of thoughts and emotions as they are formed, scientists have learned more in the past ten years about how the human brain works than in all of previous history. Indeed, their knowledge is doubling every ten years.

Thousands of scientists from many disciplines are on an unprecedented odyssey to explore the universe's most complex matter. In many respects it is the highest quest that humans are capable of—the brain understanding itself. So alluring is this goal that the brightest minds are being attracted to it. The Society for Neuroscience, for example, was formed in 1969 with 500 members. Today its membership is close to 30,000.

The researchers are on an exciting adventure that has awed the scientists themselves. It also awes anyone who comes in contact with their work, as it did me and many readers of my *Tribune* series, which put together the amazing story of their breakthroughs. The most exciting thing for me was the realization that this new understanding has the power to free the brain to achieve its maximum potential, to reach undreamed of heights.

But the research has revealed a dark side—how easy it is for things to go wrong with the brain. This kind of information is necessary to undo much of the unintentional damage inflicted on the brains of those young people who grow up in impoverished, violence-wracked environments without any emotional support to buffer their psychological trauma.

The first three years of a child's life are critically important to brain development. Unfortunately, for a growing number of children, the period from birth to age three has become a mental wasteland that can sustain only the gnarled roots of violent behavior. Society needs to focus on this period if it is to do something about the increasing rates of violent and criminal acts.

So new is most of the information about the brain, and so swiftly is it being discovered, that the real message of the latest advances is often hidden under a mountain of scientific data.

In search of answers to the editor's question, I spent the next two years on my own odyssey to find out what the new research meant. As each scientist told me his or her story, he or she would invariably point me to other researchers who were working on related issues. In all, I interviewed more than 300 researchers from many countries and read mounds of scientific articles. Their stories were like individual pieces of a grand puzzle. When I put them all together they revealed the scope of the revolution under way to fathom the workings of the brain.

By synthesizing these findings, I was able to elucidate the new understanding of the brain, which challenges deeply held beliefs about education, child-rearing, criminal behavior, treatment of mental diseases once thought untreatable, and memory enhancement. This new knowledge has profound implications for public policy issues such as child development, mental health, the criminal justice system, and self-improvement.

In essence, the information provides the best answer yet to the question: How can a newborn with thousands of muscles, scores of organ systems, 100 billion brain cells, and trillions of connections between these cells ever figure out how to get them all working together to produce consciousness, reason, memory, language, and a seemingly infinite array of adaptations to any environment in which it finds itself?

How does that happen? One of the first answers I got to that question came from the University of Chicago's Peter Huttenlocher and hit me like a bolt of lightning. I had been covering brain research for more than twenty years, and in all that time scientists could look at the three-pound organ only from the outside. Huttenlocher's remarks made me suddenly realize that scientists were actually prying open the "black box."

One of the joys of being a science writer is discovering new ideas. What Dr. Huttenlocher told me registered off the "joy" scale. He was counting synapses, the telephone lines that enable brain cells to communicate with each other. Synapses are so small and so numerous that they had previously defied a scientific census. He found that at times synapses were forming in a new brain at the incredible rate of 3 billion a second.

At eight months, a baby's brain has about 1,000 trillion connections. After that they gradually begin to decline. Half the connections die off by age ten or so, leaving about 500 trillion that last through most of life. What was the reason for this vast overproduction and pruning away? It turns out that twice as many connections are made in order to guarantee that a newborn will have sufficient extra wiring to be able to receive input from any environment it is born into—whether Borneo or Boston—and to adapt to the food, language, and culture. Other scientists have shown that synapses come and go with mental stimulation and that they are the key to brain power.

The first section of the book describes the new understanding of how the brain gets built—how it uses the outside world to shape and reshape itself, and how it undergoes crucial phases of development during which the presence or absence of appropriate stimulation can have lifelong effects, both good and bad.

This tremendous flexibility, which enables the brain to constantly undergo physical and chemical changes as it responds to its environment, is called plasticity. It is a startling departure from the old concept of the brain as a self-contained, hard-wired unit that learns from a preset, unchangeable set of rules.

Such information provides new insights into human behavior and a clearer understanding of how nature and nurture interact to determine the kind of people our children become.

The second section describes how the brain gets damaged, whether from environmental threats and stresses, or from alcohol, stroke, or head trauma. The focus is on the work of scientists who are at the cutting edge of research into the biology of violence. They are on a pioneering expedition into the uncharted area that is called the genetics of behavior.

Studies pinpointing how aggression is triggered in the brain and how it might be prevented offer hope that the nation's growing epidemic of violence might someday be slowed and even halted by treating its root causes. At the same time, a better understanding of how alcohol induces craving opens the door to new treatments and prevention of alcohol and drug addiction. Likewise, much of the damage caused by stroke or traumatic brain injury now appears to be preventable.

The third section of the book describes how the brain heals itself. It is a guided tour through the corridors of the mind where scientists are searching for the hormones and chemical messengers that keep the brain young and healthy.

No other time in history has offered such promise because no other time has had the technology to probe the brain's mysteries. These powerful tools include modern genetics, which can identify faulty genes; imaging devices, which can see how faulty genes misbehave biologically and emotionally in the living brain; molecular biology, which can correct chemical mistakes; and genetic engineering, which can change the course of a disease.

Using these new tools, scientists hope to prevent and even reverse the ravages of Alzheimer's disease, memory loss, and other mental problems that rob too many people of the very thing that makes them human. Memory pills, once the province of science fiction, are nearing reality, as are hormone replacment techniques and brain tissue transplants. The most important discovery for me was that the brain gets better and better through exercise but "rusts" with disuse. It is the ultimate use-it-or-lose-it machine, placing the ability to build brain power squarely into the hands of each one of us.

—Ronald Kotulak

Part One HOW THE BRAIN GETS BUILT

1 Revealing the Secrets of the Brain

Seeing stars, it dreams of eternity. Hearing birds, it makes music. Smelling flowers, it is enraptured. Touching tools, it transforms the earth. But deprived of these sensory experiences, the human brain withers and dies.

Scientists long have wondered how the brain can do all the things that make one person a poet, another a builder or musician, and still another a criminal or social dropout. Until recently, medical researchers never thought they could understand the brain's inner workings. They could see that a child who is loved and given stimulating experiences usually turns out to be a bright, affable person, while an abused child often becomes an abuser. But no one knew what happened inside the brain that made one person a success and another antisocial.

Researchers were resigned to measuring what went into the brain and studying what came out. The brain simply was considered the "black box." But now many secrets are being revealed.

Two of the most surprising and profound discoveries are that the brain uses the outside world to shape itself and that it goes through crucial periods in which brain cells must have certain kinds of stimulation to develop such powers as vision, language, smell, muscle control, and reasoning.

The new discoveries are overturning the old concept of a static brain, a self-contained unit that slowly begins the process of learning from a preset, unchangeable set of rules, like a tape recorder that stores whatever words it happens to hear.

Now, thanks to a recent revolution in molecular biology and new imaging techniques, researchers believe that genes, the chemical blueprints of life, establish the framework of the brain, but then the environment takes over and provides the customized finishing touches. They work in tandem. The genes provide the building blocks, and the environment acts like an on-the-job foreman, providing instructions for final construction.

These discoveries are changing the way we think about thinking and are illuminating the biological causes of behavior.

"Within a broad range set by one's genes, there is now increasing understanding that the environment can affect where you are within that range," said Dr. Frederick Goodwin, former director of the National Institute of Mental Health. "You can't make a 70 IQ person into a 120 IQ person, but you can change their IQ measure in different ways, perhaps as much as 20 points up or down, based on their environment."

The discovery that the outside world is indeed the brain's real food is intriguing. The brain gobbles up its external environment in bits and chunks through its sensory system: vision, hearing, smell, touch, and taste. Then the digested world is reassembled in the form of trillions of connections between brain cells that are constantly growing or dying, or becoming stronger or weaker, depending on the richness of the banquet.

"Just as the digestive system can adapt to many types of diet, the brain adapts to many types of experiences," says Felton Earls, professor of human behavior and development at the Harvard School of Public Health and professor of child psychiatry at the Harvard Medical School.

How a newborn learns either English or Hindi or adjusts to being raised in Sweden or Ghana or to eating a diet of beef and potatoes or raw fish and seaweed are all due to the brain's great flexibility.

"All infants require milk before they can eat solids," Earls said. "Is there an equivalent state of affairs for the brain? The answer is clearly an affirmative one. It requires stimulation: touch, holding, sound, and vision."

Several recent animal experiments have demonstrated how brain cells can rearrange their 500 trillion or so connections in response to the stimuli they are being fed.

- Vision. Magrinka Sur of the Massachusetts Institute of Technology converted brain cells that interpret sounds into cells that can process visual images by reconnecting them to the stimuli coming in through the eyes. The experiment demonstrated the interchangeability of brain cells in early development.

- Touch. When monkeys were allowed to use only one finger to perform a task, neuroscientist Michael Merzenich of the University of California at San Francisco found that the brain cells that had been committed to the now-useless fingers switched their function to other parts of the hand. Amazingly, even mature brain cells can perform totally new tasks.

- Smell. Eager to learn from the moment of birth, an infant first bonds with its mother through its sense of smell. Michael Leon of the University of Southern California discovered that within seconds of the first time a newborn smells its mother's body, indelible networks rapidly form in its brain.

- Sound. Without proper stimulation, the connections that allow brain cells to process sound, and thus, language, become scrambled. They don't form the neat columns of cells that are so characteristic of the brain's architecture. According to Martha Pierson of the Baylor College of Medicine in Houston, such scrambling may cause childhood seizures, epilepsy, and language disorders. Pierson's remarkable experiment showed how experience, or the lack of it, can physically change the brain and cause mental disorders.

"It's just phenomenal how much experience determines how our brains get put together," Pierson, a neurobiologist, said. "If you fail to learn the proper fundamentals at an early age, then you are in big trouble. You can't suddenly learn to learn when you haven't first laid down the basic brain wiring. . . . That's why early education is so important, why Head Start is so important," she said, referring to the federally funded program for preschoolers.

Essentially, a human comes equipped with a brain for all places and all ages. It takes in stride TV, the transition from horses and buggies to jets, and moon travel in a single lifetime.

But what the brain can do depends on whether or not it is used. It is the ultimate use-it-or-lose-it machine, and it is eager to learn new skills. The ability to form abstract thoughts, for instance, is now seen as a consequence of the brain's learning to read.

"A thousand years ago in medieval England most people did not think abstractly," said Dr. Bruce Perry, a Baylor College of Medicine neuropsychiatrist. "The majority of people viewed the world very concretely. When we look back now and think about how superstitious they were and all that kind of stuff, it's not that dissimilar from the way eight- or nine-year-old children today think about things and view the world.

"In the same way that we evolved a certain cognitive abstract capability as a function of our capacity to read, there is every reason to believe that there are other untapped abstract capabilities of our brains that are not being developed by our traditional educational system."

In their quest to learn how the brain works, scientists have found that the three-pound, walnut-shaped mass of gray matter goes through four major structural changes: in fetal development, after birth, between four and twelve, and the years thereafter.

Starting from a few cells at the tip of an embryo, brain cells multiply at an astounding rate: About 200 billion are created in several months. Their job is to get in touch with the body that is developing around them and they compete to succeed. Half of the brain cells die off by the twentieth week of fetal life because they fail to connect to some part of the awakening body.

This overproduction of brain cells is important: It is evolution's way of making sure there are enough cells to handle the development of new skills, just as brain cells did in past generations to develop upright walking and language.

During the winnowing-down phase, the brain is organized into more than forty different physical "maps," which broadly govern such things as vision, language, muscle movement, and hearing. How these maps are organized is influenced by electrochemical signals coming into the brain from all parts of the body, and by hormones. Sex hormones are especially potent because they can physically shape a male or female brain and influence its skills, favoring such things as language in females and spatial abilities—mathematical concepts, for example—in males.

Alcohol and drug abuse can interfere with growing brain cells, jamming their genetic performance and increasing the risk of mental disorders. Alcohol-induced birth defects, for instance, are the leading known preventable cause of mental retardation in the United States, affecting 1 in every 800 to 1,500 newborns.

Long thought to be a clean slate to which information could be added at any time, the brain is now seen as a super-sponge that is most absorbent from birth to about the age of twelve. Thus, the brain can reorganize itself with particular ease early in life during crucial learning periods, when connections between brain cells are being made and broken down at an enormous rate. Information flows easily into the brain through "windows" that are open for only a short duration. These windows of development occur in phases from birth to age twelve when the brain is most actively learning from its environment. It is during this period, and especially the first three years, that the foundations for thinking, language, vision, attitudes, aptitudes, and other characteristics are laid down. Then the windows close, and much of the fundamental architecture of the brain is completed.

"A kind of irreversibility sets in," Harvard's Earls said. "There is this shaping process that goes on early, and then at the end of this process, be that age two, three or four, you have essentially designed a brain that probably is not going to change very much more."

That's not to say that all is lost if this early learning period is not optimized. Using the tools left over from shaping brain cells and their connections, the brain gives its owner a big second chance which runs to about age twelve. Even after that the brain never stops learning. There is, however, a price to pay. Instead of being easy, learning becomes harder later on, as any adult who has tried to learn a foreign language knows. For a child, foreign languages are picked up easily.

The brain learns and remembers throughout life by employing the same processes it uses to shape itself in the first place: constantly changing its network of trillions of connections between cells as a result of stimuli from its environment.

One of the most striking examples of this ability to change was shown recently by Bruce McEwen of Rockefeller University. During the four-day reproductive cycle of a female rat, he found, new connections are created and old ones are destroyed as hormones prepare their brains for pregnancy and later to care for their pups.

"People hear that and say, 'My God, that's amazing!' and these are neuroscientists," he said. "A lot of people are surprised at the rapidity with which connections can be made and broken down in the brain. It especially comes as a big surprise to people who take a more psychological view and separate the mind from the brain. They are part and parcel of the same thing. It doesn't degrade your ability to talk about higher cognitive function [when you] realize that there's a brain under there that's doing the work."

Surprisingly, almost anything can cause physical changes in the brain: Sounds, sights, smells, touch—like little carpenters—all can quickly change the architecture of the brain, and sometimes they can turn into vandals.

"The new thing is that the brain is very dynamic," said Dr. Robert Post, chief of the National Institute of Mental Health's biological psychiatry branch. "At any point in this process you have all these potentials for either good or bad stimulation to get in there and set the microstructure of the brain."

Post and his colleagues were startled to find that outside stimulation can permanently alter the function of brain cell genes. Stress and drugs like cocaine, for instance, can produce biochemical changes that directly affect the function of some key brain-cell genes, in effect laying down permanent, maladaptive behavior patterns.

Faced with the new evidence about how the brain develops and functions, many scientists are concluding that society is wasting a tremendous amount of the brain power of its young, and creating a lot of unnecessary problems—including crime, aggression, and depression—later on in their lives.

"We are underinvested in our children," said Frederick Goodwin. "We spend seven times more per capita on the elderly than we do on children. Now that we have better concepts of the plasticity of the brain, it is obvious we are wasting a tremendous resource."

Understanding the role of the environment in altering brain plasticity has opened the door to prevention, he said. "The question now is if we can identify the kids who are the most vulnerable to being damaged by their environment and get the plasticity of the nervous system working for us to prevent such damage," Goodwin said.

Recent research shows that proper stimulation affects such brain functions as:

- Language. Children whose mothers talk to them frequently have better language skills than do the children of mothers who seldom talk to them. After about age twelve the ability to learn new languages declines rapidly.

- Vision. Lack of visual stimulation at birth will cause those brain cells designed to interpret vision to dry up or be diverted to other tasks, making perfectly healthy eyes permanently unable to see. This discovery has saved the sight of thousands of infants born with vision-blocking cataracts, which are now removed as quickly after birth as possible.

- Brain power. Mice and rats raised in enriched environments, with toys and playmates, have billions more connections between brain cells than similar mice raised alone in empty cages. Pioneering studies also show that the IQs of children born into poverty, or of those who were premature at birth, can be significantly raised by exposure to toys, words, proper parenting, and other stimuli.

- Aggression. Early exposure to violence, stress, and other environmental pressures can cause the brain to run on a fast track, increasing the risk of impulsive actions and high blood pressure.

- Emotions. Animals exposed to unpredictable stresses while still in the womb develop anxious personalities. After birth, a little extra

THE BRAIN: HOW IT WORKS AND DEVELOPS

New discoveries are changing old concepts of how the brain develops and works. Two of the most surprising discoveries indicate that the brain uses the outside world to shape itself, and that it goes through critical periods in which brain cells require specific types of stimulation to develop such powers as vision, language, smell, muscle control, and reasoning. A related discovery is that the brain has the ability to change rapidly as it physically reshapes itself into a kind of biological map of the outside world. Researchers now believe that genes establish the framework of the brain, but the external environment provides the customized finishing touches.

Mapping the cerebrum Areas and their known functions

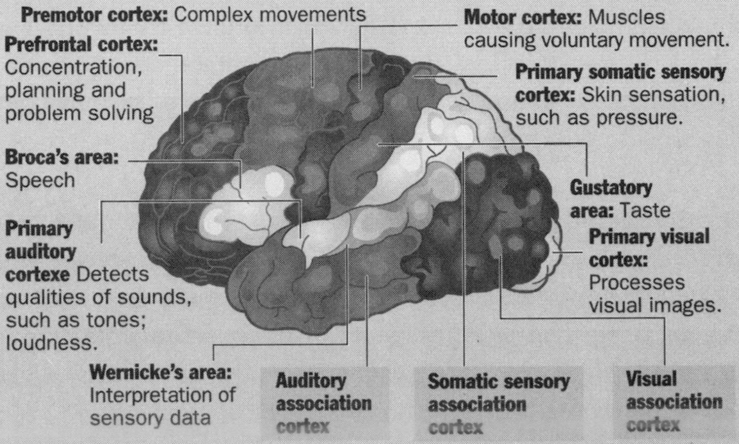

THE GROWING BRAIN

Major structural developments

- **Fetal development:** Billions of brain cells are formed in the first months of fetal life. Half of them die as hormones and other stimuli eliminate and organize them to form the brain's basic scaffolding, e.g. male or female.
- **After birth:** Trillions of brain cell connections are established and form the brain's physical "maps" that govern such things as vision, language, and hearing.
- **Age 4 to 10:** New learning reorganizes and reinforces connections between brain cells. New connections are formed as new things are learned.
- **After age 10:** Still able to undergo physical changes, the brain learns and remembers throughout life.

ASSOCIATION AREAS:

Areas that further interpret information received by primary areas.

Example: The primary auditory cortex detects simple sounds such as pitch and volume, while the auditory association cortex analyzes that information and enables recognition of whole sounds, such as spoken words.

Sources: *ABC's of the Human Body,* American Medical Association, *The Human Body.*

Chicago Tribune/Steve Little, Terry Volpp

mothering has the opposite effect, instilling them with confidence and the urge to explore.

- Touch. Premature infants whose sensory systems are activated by being held and cuddled are more mentally alert and physically stronger than those who are routinely isolated in incubators.

- Education. The best time to learn foreign languages, math, music, and other subjects is between one and about twelve years of age, yet these years are usually put on pause, given over to youngsters to "enjoy their childhood."

"The aspects of brain development most closely tied to human behavior can be affected for better or worse by the care we give our children," said Yale University neurobiologist Martha Constantine-Paton. Such knowledge provides the moral and social imperative to prevent or cure brain damage caused by the lack of proper environmental stimulation during the brain's crucial periods of development in fetal life and childhood.

"Legislative and educational efforts aimed at nurturing the developing brain through these critical periods could be instituted in the immediate future if the collective public conscience realized that the actual structure of the brain can be adversely affected by neglect," she added.

What can parents do to ensure that the brains of their children develop properly?

"If you want to significantly influence a child's ability to think and to acquire knowledge, the early childhood years arc very critical," said neurobiologist Peter Huttenlocher, whose studies helped open the door to understanding the brain's plasticity.

Rockefeller University's McEwen says: "The most important thing is to realize that the brain is growing and changing all the time. It feeds on stimulation and it is never too late to feed it."

2 Mental Workouts and Brain Power

Counting the grains of sand on Oak Street Beach would have been easier, but then University of Chicago neurobiologist Peter Huttenlocher wouldn't have discovered a key to unlocking one of the brain's deepest secrets. Peering into the lens of a new, powerful electron microscope, Huttenlocher was counting for the first time the connections between brain cells—the tiny, numerous linkages that make the brain a thinking organ.

The brain samples, which had been removed during autopsies on fetuses, deceased babies, and elderly people, contained more than 70,000 cells each, even though they were only the size of a pinhead. As he focused on one sample and then another, Huttenlocher was astonished by the sharply increasing number of brain-cell connections; he later compared the sight to watching the slow-motion frames of an explosion.

A sample from a twenty-eight-week-old fetus totaled 124 million connections between cells; the sample from a newborn, 253 million; and the sample from an eight-month-old—an amazing 572 million connections, findings that defied conventional scientific wisdom.

"It was a strange thing to see," he said. "The number of connections kept going up and up and up and then they started to go down." Brain connections, he soon learned, start to fizzle toward the end of the first year of life, stabilizing at 354 million per speck of brain tissue by age twelve.

By the time he was done with his census, Huttenlocher not only stunned neuroscientists with his demonstration of how fast the brain initially develops, he also provided a glimpse of the brain's raw power to create a powerful learning machine. "I stumbled on the whole thing," said Huttenlocher, who launched the census about a decade ago. "It was something that nobody expected. It took quite a long time until people began to accept that this really happens."

Huttenlocher's pioneering census was the first hint of something that has evolved into accepted scientific fact today: The brain is not a static organ; it is a constantly changing mass of cell connections that are deeply affected by experience and hold the key to human intelligence.

By probing further into Huttenlocher's insights, scientists have since developed measurable biological explanations for imprinting, bonding, and other critical periods of learning that parents, educators, and scientists could only guess at before.

At the University of Illinois at Champaign-Urbana, for example, William T. Greenough, a psychologist and cell biologist, found that rats raised in enriched environments with toys and other animals to play with had measurably more connections between brain cells and were better learners than those raised in less stimulating surroundings.

Scientists also are zeroing in on the idea that aggression, violence, and crime are rooted in the brain's biological reactions to violent and stressful experiences. The genesis for these insights dates back decades to when scientists, puzzled by the mysteries of learning problems and physical disabilities, began looking to the brain for some answers. No one suspected that the brain was as changeable as science now knows it to be.

The first brain cell, or neuron, is thought to have appeared in animals about 500 million years ago. Able to form flexible connections with other cells to send and receive electrochemical messages, the neuron marked a crucial leap in evolution, second only to that of the DNA molecule that appeared some three billion years earlier. Just as DNA gave birth to life in all its forms, the neuron made possible complex functions and, eventually, creative thought.

Scientists discovered that the power of a brain grows in direct relationship with the number of cells it has. "Our brains are built from the same molecular bricks as the other species, but we have a different building, a new architectural structure with more bricks," said

THE BRAIN'S COMPUTER CHIPS

● Brain cells, also known as neurons, receive, analyze, coordinate, and transmit information. The brain learns and remembers throughout life by constantly changing its network of trillions of connections between neurons as a result of stimuli from its environment. Researchers now know that some of these connections, or synapses, grow stronger with learning and weaken or disappear when not used.

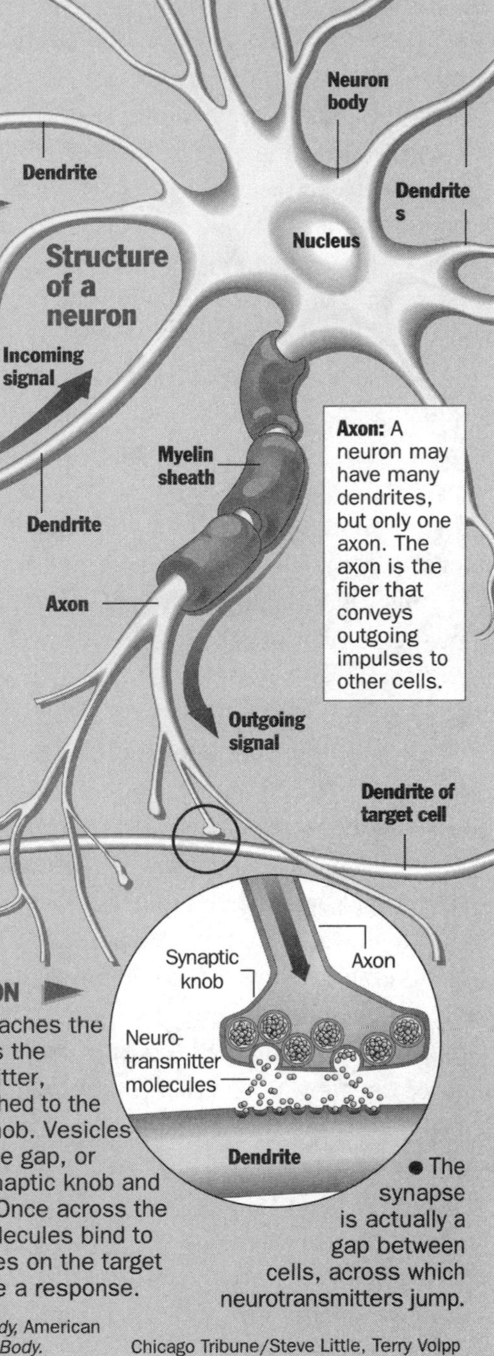

Dendrites: Fibers extending from the body of the neuron which receive incoming signals from other neurons.

MAKING THE CONNECTION

● An electrical impulse reaches the axon terminal and triggers the release of a neurotransmitter, contained in vesicles attached to the surface of the synaptic knob. Vesicles spill their contents into the gap, or synapse, between the synaptic knob and the target cell's surface. Once across the gap, neurotransmitter molecules bind to specific receptor molecules on the target cell's surface and activate a response.

● The synapse is actually a gap between cells, across which neurotransmitters jump.

Sources: *ABC's of the Human Body,* American Medical Association, *The Human Body.*

Chicago Tribune/Steve Little, Terry Volpp

neurobiologist Pasko Rakic of Yale University. "The secret is in the total number of brain cells and the number of connections between them."

A fruit fly has 100,000 brain cells, a mouse 5 million, and a monkey, man's closest relative in the animal kingdom, 10 billion. Each is equipped with the brain power it needs to live in a particular environmental niche, evolution's way of finding a home for everything.

But nature overshoots its mark with humans. In only thirty-six hours of embryonic development, the human brain makes ten times more cells than a monkey's. Having 100 billion cells puts the human brain into a class by itself. That kind of computing power allows us to leapfrog into new realms: self-awareness, language, making associations, and abstract thinking. We can remember the past and anticipate the future so as to better guide ourselves in the present.

That is what is generally called consciousness. With less capacity, other animals are locked in the present, forced to depend on instinct.

During fetal development the human brain goes wild, producing twice the number of cells it will eventually keep. Before birth about half of these cells are killed off when they fail to find a job to perform.

Brain cells compete to connect to some part of the body. Those cells that fail to hook up do not get the proper feedback, which includes chemicals that nourish and maintain brain cells. Left without such nourishment or a sense of direction, they perish.

About halfway through fetal life, when some brain cells start to die and a unique brain takes shape, sex differences and temperament begin to emerge, says Harvard University child psychiatrist Felton Earls.

After birth, another wild spurt of growth occurs as the brain races to make the connections that Huttenlocher measured between all of its cells, a profusion that happens so fast that it would be easier to count the drops of water in a rainstorm. Then another massive die-off occurs as half of the connections disappear by puberty. This time the death of connections, called synapses, is caused by a lack of interaction with the outside world. Connections that are not strengthened by stimulation from the environment die off.

Left behind in the average brain are as many as 500 trillion wiggling conduits that are ready to flash messages between brain cells. The number of connections could easily go up or down by 25 percent or more, depending upon whether a child grows up in an enriched environment or in an impoverished one.

The net effect is that the brain produces many more cells and connections than it could ever use. Both phenomena are examples of genetic frugality. Humans do not contain anywhere near enough genes to make the individual cells that make up a fully operational brain. So an overabundance of the same or similar cells and synapses are produced and then the brain has to learn how to make itself work.

The trillions of connections that survive the great die-off owe their survival in large part to what a child learns in his or her first decade.

Learning foreign languages is a clear example. Before puberty most children can easily learn a language without an accent. The excess supply of connections that are available to be called into service enables a youngster to learn the slightest nuances of sounds as they are spoken. As these connections dwindle, however, languages become harder to learn, and when they are learned they are almost always accompanied by an accent.

Language development helps illustrate how the brain is genetically programmed to respond to stimuli at certain critical periods. The sounds of words, for example, are received by receptors in the ear and converted into electrochemical signals. The signals travel along nerves to specific parts of the infant brain where they awaken cells to their potential to process language. Millions of language cells swing into action, generating new tendrils to connect with other brain cells.

Without exposure to spoken words, cells that allow the brain to construct meaningful sentences do not develop properly. They die on the vine or their function is usurped by more aggressive cells in other parts of the brain. In the process, their owner is cheated of the brain's full potential to use language, as has been shown by children who grow up alone in the wild.

Appropriate early stimulation is likewise crucial for the development of vision and other sensory functions. Scientists now believe that everything else that the brain regulates—learning, memory, emotions, physiological responses like reaction to stress and high blood pressure—are molded in early development when the brain changes the most.

In adulthood, the brain finally settles down but it is not idle. It will keep building and destroying connections and strengthening and weakening established ones for the rest of its life as it adjusts to the continuous changes in its environment.

"The fetal brain and the developing brain are very different structures from the adult brain, and the developing brain of a baby is really not a small version of an adult brain," said Carla J. Shatz, a neurobiologist at the University of California at Berkeley. "That takes people by surprise," she said. "They always think the brain just gets bigger."

William Greenough found that he could quickly increase the number of connections in animal brains by 25 percent simply by exposing the animals to an enriched environment. "What we know from animals suggests that the harder you use your brain, whether it's thinking or exercising, the more in shape it's going to be," Greenough said.

And that's exactly what seems to be happening in the human brain. Scientists at the University of California at Los Angeles recently found in autopsy studies that the brains of university graduates who remained mentally active had up to 40 percent more connections than the brains of high school dropouts.

"This is the human equivalent of the animals exposed to enriched environments having smarter brains," said UCLA neuroscientist Bob Jacobs. "We found that as you go up the educational ladder there is a dramatic increase in dendritic material."

Dendrites sprout from brain cells like tree branches. Studded along the branches like leaves are junctions called synapses, which connect to other brain cells. Just as leaves receive light from the sun to enable a tree to grow, synapses receive information from other brain cells to increase the brain's power to think.

But education alone is no guarantee of a better brain, the UCLA scientists found. Unless the brain is continuously challenged, it loses some of the connections that grew out of a college experience. The brains of university graduates who led mentally inactive lives had fewer connections than those of graduates who never stopped letting the light in. "The bottom line is that you have to use it or you lose it," Jacobs said.

An unanticipated bonus of an educated brain is that it may be better protected against Alzheimer's disease, scientists are finding. With more connections serving as front-line defenses, "educated" brains can better withstand the destructive attacks of Alzheimer's.

The power of experience to shape the brain struck like a thunderbolt in the 1960s and 1970s, and it took many years for scientists to believe it. It started with a landmark series of experiments at Harvard Medical School in which Torsten Wiesel and David Hubel sewed shut one eye of newborn kittens to test the effects of sensory deprivation. When the eyes that had been stitched closed were opened a few weeks later, the eyes were not able to see. But, surprisingly, the eyes that had remained open could actually see better than normal eyes.

Something strange had happened. The brain cells that normally would have been committed to processing visual stimulation from the closed eye had failed to learn that task. But they had gone off to help the other eye, and no amount of visual coaxing could get them back.

Wiesel and Hubel had made two important discoveries for which they won a Nobel Prize. They showed that sensory experience is essential for teaching brain cells their jobs, and after a certain critical period, brain cells lose the opportunity to learn those jobs.

That failure to learn is well known in real life. Even if a person's brain is perfect, if it does not process visual experiences by thc age of two, the person will not be able to see, and if it does not hear words by age ten, the person will never learn a language.

"These are very important insights," said Wiesel, who is now president of Rockefeller University. "There is a very important time in a child's life, beginning at birth, when he should be living in an enriched environment—visual, auditory, language, and so on—because that lays the foundation for development later in life."

What was true for the cats was also true for humans. Babies born with cataracts, clouded lenses that prevented visual experiences from reaching the brain, also grew up blind because the brain cells that would normally process vision were called to duty elsewhere.

Based on the discoveries of Wiesel and Hubel, surgeons abandoned their practice of waiting until a child was several years old before removing cataracts. They began removing cataracts early, thereby preventing blindness by allowing visual experiences to reach the brain.

On a more fundamental level, Wiesel and Hubel showed what brain cells really are up to. The brain, it seems, is the ultimate reductionist. It reduces the world to its elemental parts—photons of light, molecules of smell, sound waves, vibrations of touch—which send electrochemical signals to individual brain cells that store information about lines, movements, colors, smells, and other sensory inputs.

Wiesel and Hubel discovered this when they implanted tiny electrodes into cells of the visual cortex of cats and monkeys. One set of cells recognized perpendicular lines. Another set of cells next to them recognized only lines slanted at a one o'clock angle; the next set recognized lines at two o'clock, and so on.

As a data-storage strategy, it was brilliant. A countless number of images could be disassembled into their parts and stored in specialized brain cells. One cell then could be used many times to recall similar lines found in a building, book, or car. With hundreds of trillions of connections to work with, the brain can establish flexible circuits between groups of cells that capture the various parts of an experience, whether it be a face, sunset, smell, or meal.

Thus, there are no pictures stored in the brain, as was once thought. There are patterns of connections, as changeable as they are numerous, that, when triggered, can reassemble the molecular parts that make up a memory. Each brain cell has the capacity to store fragments of many memories, ready to be called up when a particular network of connections is activated.

How many connections are kept depends on the learning experiences the brain is exposed to. The evidence indicates that the more connections you have, the smarter you are. "We do waste a lot of brain power because we're not quite sure how to exploit our brains to the maximum," said Lawrence Garey of the University of London. "We're just learning how to do that."

PATH OF THE MIND'S EYE

The brain and the eye work together to process images by converting them into specialized electrochemical signals, then storing these components in the visual cortex for future recall.

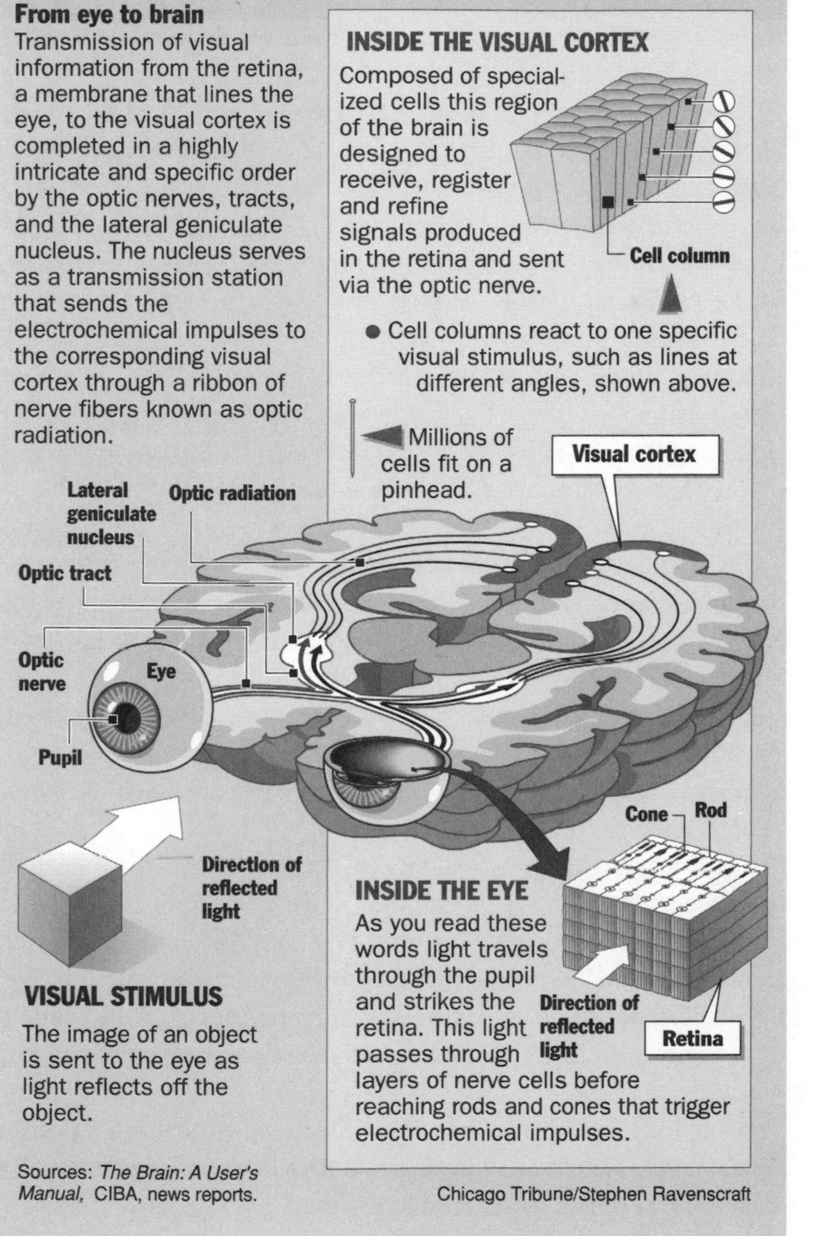

Another important lesson scientists have learned is that words can be just as powerful as drugs in correcting errant brain pathways that are causing some mental diseases.

Using high-tech imaging devices that can "see" the living brain processing thoughts, scientists at the University of California at Los Angeles showed for the first time that behavior therapy produced the same kinds of physical changes in the brain as psychoactive drugs.

Experts hope that the findings will shed new light on the process by which psychotherapy, including the so-called "talking cure," can physically change the brain and that they will lead to better techniques for treating mental diseases. "Anytime you have a change in behavior you have a change in the brain," said UCLA psychiatrist Dr. Lewis Baxter. "Behavior therapy and drugs appear to rearrange brain circuitry in the same way."

Baxter and his colleagues studied the brains of obsessive-compulsive patients with PET (positron emission tomography) scans, a technique that measures the activity of cells in different areas of the brain. They found that an area called the caudate nucleus was overactive in these patients.

The caudate nucleus acts as a gatekeeper that prevents unwanted thoughts from establishing self-reinforcing circuits in the brain. Like a record stuck in the same groove, unwanted thoughts keep repeating themselves and drive compulsive behavior.

In some cases patients are afraid of dirty objects and repeatedly wash their hands. Others may be fearful of violence and check their door locks hundreds of times a day. Obsessive-compulsive disorder affects 2 percent of the population.

Fluoxetine (Prozac) is highly effective in curbing unwanted thoughts through a mechanism that is little understood, but which involves raising the level of an important brain chemical messenger called serotonin. Serotonin plays a key role in controlling impulses. So when serotonin is low, impulses such as repeated hand washing or constantly checking to make sure the stove is off can keep popping out.

Behavior therapy, which is also effective in breaking unwanted habits, involves gradual exposure to a fear-triggering agent, such as dirt, and teaching a patient not to respond to compulsive urges.

Although the two treatments appear to be highly dissimilar, PET scans showed that they produced identical changes in calming down caudate nucleus activity in both groups. Both therapies appear to correct the abnormal circuits causing unwanted thoughts by changing connections between brain cells in the caudate nucleus, Baxter said.

His experiment is one of the most dramatic demonstrations of the power that words have to physically change the brain.

3 How the Brain Learns to Talk

The words came slowly, and at first they were a little difficult to understand, but the children who spoke them had crossed into a new world and they beamed. For those who witnessed the scene, there was no mistaking that a medical miracle was taking place at the Indiana University School of Medicine.

Children who were born deaf and had never learned to speak were hearing and using spoken language for the first time with the aid of surgically implanted devices that sent sounds into their brains.

"I couldn't believe it," said Dr. Mary Joe Osberger, director of research in the university's department of otolaryngology. "We thought we might be putting sounds like pops, buzzes, and clicks into their brains. But their brains heard them as words, and as words they came out."

The experiment is just one example of science's quest for insights into the relationship between the brain and language, the "gift" that separates humans from all other species and enables humans to bring form to their imaginations, ideas, and needs.

The implications of the inquiries are breathtaking. If science can develop a better understanding of language development problems, experts believe that strides can be made in dealing with a host of other emotional, social, and behavioral problems.

Armed with sophisticated new technology like the artificial ear and imaging devices that can peer inside working brains, scientists hope to gain insights into how language development affects reading problems, learning disabilities, and problem-solving.

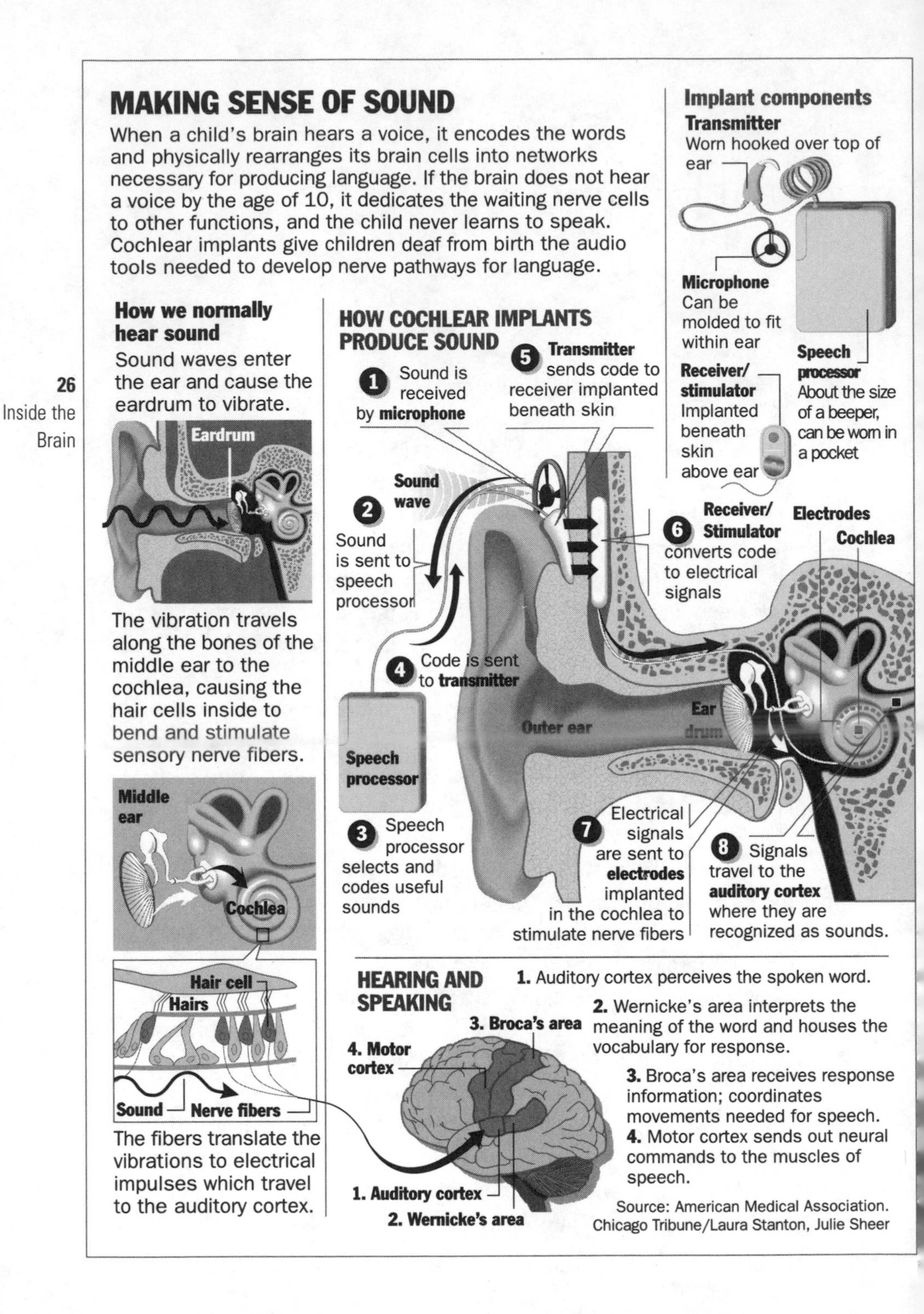
MAKING SENSE OF SOUND
When a child's brain hears a voice, it encodes the words and physically rearranges its brain cells into networks necessary for producing language. If the brain does not hear a voice by the age of 10, it dedicates the waiting nerve cells to other functions, and the child never learns to speak. Cochlear implants give children deaf from birth the audio tools needed to develop nerve pathways for language.
Implant components
Transmitter
Worn hooked over top of ear
Microphone
Can be molded to fit within ear
Speech processor
About the size of a beeper, can be worn in a pocket
Receiver/ stimulator
Implanted beneath skin above ear
How we normally hear sound
Sound waves enter the ear and cause the eardrum to vibrate.
Eardrum
The vibration travels along the bones of the middle ear to the cochlea, causing the hair cells inside to bend and stimulate sensory nerve fibers.
Middle ear
Cochlea
Hair cell
Hairs
Sound
Nerve fibers
The fibers translate the vibrations to electrical impulses which travel to the auditory cortex.
HOW COCHLEAR IMPLANTS PRODUCE SOUND
1 Sound is received by microphone
2 Sound is sent to speech processor
Sound wave
3 Speech processor selects and codes useful sounds
Speech processor
4 Code is sent to transmitter
5 Transmitter sends code to receiver implanted beneath skin
6 Receiver/ Stimulator converts code to electrical signals
Electrodes
Cochlea
Outer ear
Ear drum
7 Electrical signals are sent to electrodes implanted in the cochlea to stimulate nerve fibers
8 Signals travel to the auditory cortex where they are recognized as sounds.
HEARING AND SPEAKING
1. Auditory cortex perceives the spoken word.
2. Wernicke's area interprets the meaning of the word and houses the vocabulary for response.
3. Broca's area receives response information; coordinates movements needed for speech.
4. Motor cortex sends out neural commands to the muscles of speech.
3. Broca's area
4. Motor cortex
1. Auditory cortex
2. Wernicke's area
Source: American Medical Association.
Chicago Tribune/Laura Stanton, Julie Sheer

Despite the pervasive role of language in society, science knew little of how humans developed language until recently. Up to the time they conducted the experiment at Indiana University, for example, scientists had no idea that deaf people could process even simple sounds into language—if the sounds could find their way into the brain.

"These are children who are not partially deaf, they are totally deaf," Osberger said. "Until now, their ability to speak was zero, no matter how intensely they were trained."

In a child who is born deaf, the 50,000 nerve pathways that normally would carry sound messages from the ears to the brain are silent. The sound of the human voice, so essential for brain cells to learn language, can't get through and the cells wait in vain. Finally, as the infant grows older, brain cells can wait no longer and begin looking for other signals to process, such as those from visual stimuli.

The critical period for learning a spoken language is totally lost by about age ten. Children who grow up alone in the wild, never hearing another human, cannot learn to speak if they are introduced to civilization after that deadline.

By using an artificial ear called a cochlear implant, though, the scientists at Indiana University and several other medical centers activated some of those disconnected nerve pathways, enabling rough sounds of the human voice to reach the brain.

"The first thing the implant does is to start changing their brains," said Michael Merzenich of the University of California at San Francisco. "The fact that the implants work at all is amazing, but clearly they are seeing reasonable results. It's miraculous, actually."

So eager are brain cells in the auditory cortex to do their jobs that they jump at the meager sounds coming through the cochlear implant, using them to construct a vocabulary, grammar, and syntax—the rules of language.

From what scientists at Indiana and other research centers around the country now know, language appears to have been acquired late in genetic evolution. It is so new that it acts like a guest, not yet claiming a permanent position in the brain as do vision, smell, or hearing.

That language can be located in totally different areas of the brain can be seen in right- or left-handed people. While 90 percent of right-handers process speech and language in the left side of their brains, for instance, 30 percent of left-handers process language in the right side of their brains.

"Language is a very recent phenomenon," said Elizabeth Bates of the University of California at San Diego. "The odds are very good that we built it out of old stuff [in the brain] that was not originally designed to process language."

It's like building a computer out of old parts in a garage, she said. The old parts continue to do the jobs they did before—perception, memory, categorization—but now they are running at a higher speed with new computational properties.

The symbols used for language seem too simple. From a small number of sounds—vowels and consonants—a large number of words can be formed to create an inexhaustible number of sentences, each with a different meaning.

A child can learn a language merely by hearing it, thereby sharing the most complex ideas and emotions of a culture without ever having to learn to read or write. Language has become so important to humans that the brain will not let it go. Unlike vision or touch, which stay in specific areas, language can shift to different cells at opposite sides of the brain when need be.

Children who suffer major damage to the left side of the brain from accidents or disease, for instance, can acquire language using the right side of their brains. They don't need special teaching. After about age seven, however, that tremendous flexibility to bounce language around the brain diminishes. Relearning a language as a result of brain damage then becomes more difficult.

Why language prefers the left side of the brain has long puzzled scientists. Paula Tallal, co-director of the Center for Molecular and Behavioral Neuroscience at Rutgers University, believes she has found the answer.

Using brain-imaging technology to study children with a common language disorder, Tallal found that the left side of the brain processes information faster than the right side, a skill that is important for separating the sounds of speech into distinct units.

When the brain is unable to process sounds quickly, its ability to learn language is seriously impaired. An estimated 10 percent of children have just such a language disorder, called developmental language disability. "They hear and see perfectly well, but there is a problem with the way their brains process sensory information. It's as if the brain is running in slow motion," she said.

Whereas a normal six- to eight-year-old can distinguish between "ba" and "da" in 8-thousandths of a second, children with the slow-motion brains take 300-thousandths of a second.

"These kids are in big trouble in the early stages of learning to talk," Tallal said. "If you're having trouble distinguishing speech sounds, that's going to have an impact on how you learn grammar, on how you eventually learn to read, on all sorts of things."

When these children listened to language that was slowed down to stretch the intervals between the sounds of vowels and consonants, they had no trouble understanding speech that they could not comprehend at normal speeds. Discovering that the problem involves a slower processing of information inside the brain, and not a problem with the ears, eyes, or vocal cords, has opened the door to new therapies.

Correcting language disorders is vitally important. Children who do not develop normal language at the expected age are at high risk for all kinds of problems—academic, social, behavioral—that had not been previously linked to language impairment.

Studies show that a high percentage of children in psychiatric and child-guidance clinics, or who are being seen by social workers, have problems with delayed language development.

Using such brain-imaging devices as MRI (magnetic resonance imaging), which provides detailed views of brain structure, and EEGs (electroencephalographs), which outline localized brain activity, researchers have found that children with normal language skills have lopsided brains—the left side is bigger and more active than the right side. Such lopsidedness shows how the brain has partitioned itself to handle specialized jobs—the right side preferring music and orienting its owner in space, the left processing language, math, and logic.

Studies by Tallal and others found that the brains of children with language disorders had balanced brains. Both sides, the right and left hemispheres, were of equal size and activity. Having both sides equally active meant that the left hemisphere was underpowered:

It was not fast enough to adequately process the rapid staccato of speech.

For these children, the sounds of speech race through their brains like a rapid stream of water, instead of the individual drops that make up the sounds that the brain of a normal child perceives.

The left and right sides of the brain appear to become specialized during fetal development. The right hemisphere grows faster and favors more primitive characteristics, like emotion. The left hemisphere starts growing later and is in charge of newer acquisitions, such as language.

Scientists believe that all fetuses start off as female. Only when the male sex hormone, testosterone, kicks in do male features, such as a penis and larger muscles, develop. The same sex hormone exerts a profound influence on the brain, physically shaping it into a male brain.

"So you have to make a boy brain out of a girl brain at about the time in prenatal development when the two hemispheres have their greatest difference in growth," said Judy L. Lauter, an expert on communications disorders at the University of Oklahoma Health Science Center.

During this crucial period, many things can go wrong, especially with the late-blooming left hemisphere, Lauter said. These problems show up later as an increased rate of language problems in boys such as stuttering, dyslexia, and language delays.

Rutgers' Tallal agrees. Studies show that language problems in children are associated with stressful pregnancies, she said, explaining that not only can the sex hormone go awry during this time, but other compounds, such as stress hormones, can be raised to abnormal levels.

Tallal continued: "Having a very stressful pregnancy is highly correlated with the failure to show the expected structural lateralization [left and right hemisphere differences] in the brain."

The new findings emphasize the importance of reducing stress during pregnancy, say the researchers. But the discoveries also make it possible to diagnose infants who are headed for language problems long before they can even speak and to develop teaching programs to coax the brain into learning appropriate language skills.

"It's possible to find out how the brain works best and to head off future problems by designing ways for a child to learn language that are the most natural for them," Lauter said.

Scientists are looking for more efficient ways to learn language by studying how birds learn new songs every year. Songbirds follow a pattern of brain development similar to humans and they may help unravel the biology of human speech.

Both the male and female zebra finch, for instance, are born with the same brain structure. But an extra shot of hormone during development enables the song-producing part of the male brain to grow, while a similar area in females shrinks. As a result, only male finches sing. But when females are given a similar hormone boost early in development, their brains change to the male pattern and they are able to sing.

During human fetal development, brain cells are created and assigned general jobs. After birth, a second wave of structural changes occurs as an enormous number of connections, called synapses, are made between brain cells. Between birth and about eight months of age, the number of connections skyrockets from about 50 trillion to 1,000 trillion. This vast overproduction of connections, enough to ensure that the brain can adapt to any environment it finds itself in, then starts to dwindle. Connections not reinforced by what the baby is experiencing in its world—voices, sights, smells, touch—shrink and perish. Left behind are brain cells that form "maps"—a kind of biological integrated circuit—of those experiences.

How the brain puts its early learning capacity to use to store words was discovered by psychologist Janellen Huttenlocher of the University of Chicago. In a pioneering study that struck down the old notion that some children learn words faster than others because of an inborn capacity, Huttenlocher showed that when socioeconomic factors were equal, babies whose mothers talked to them more had a bigger vocabulary. At twenty months, babies of talkative mothers knew 131 more words than infants of less talkative moms, and at twenty-four months the difference was 295 words.

The babies were listening. Although it may not seem obvious, the vocabulary they are exposed to makes an impression on their brains. They are learning words faster than previously thought.

"This comes as a shock to many people, but it shouldn't," Huttenlocher said. "The prevailing wisdom had placed the mastery of vocabulary into the child as an innate capacity, and that's not true. It has to be learned."

Can TV substitute for a mother's talk? No, Huttenlocher said. "Mothers who talked a lot used the same kinds of words as mothers who talked very little. Mothers tend to talk in very short sentences. They describe the here and now. They point to the things they are talking about and rarely mention objects that aren't around. Early TV watching is words without content."

A third critical restructuring of the brain takes place between the ages of four and twelve. Educators suspected that something dramatic was happening during this period because of the surge in learning that takes place then. But pediatric neurologist Harry Chugani of Wayne State University in Detroit was the first to see it happening inside the brain. Using PET (positron emission tomography) scans to follow the brain's consumption of sugar, the energy that cells use to carry out their work, he measured the activity level of brains at all ages, from infancy to old age.

There was a big energy spurt between the ages of four and ten, when the brain seemed to glow like a nuclear reactor, pulsating at levels 225 percent higher than adult brains.

What was happening? Was the brain burning up or melting down? Neither, it turned out. It is a time when the brain is deciding whether to keep or eliminate connections. And in the process of keeping connections, the brain eagerly seeks information from the senses.

Things that a child experiences become part of his mental architecture, laid down in the connections that are retained. There is room for a virtually unlimited number of memories. Connections that are not reinforced by stimuli from the outside world are pruned away, dead branches that no longer flower.

Beyond the age of about twelve, when the brain's "maps" have been made, learning a language becomes more difficult. It involves the building of new connections and the tearing apart of old ones.

"Who's the idiot who decided that youngsters should learn foreign languages in high school?" Chugani asks. "We're not paying attention to the biological principles of education. The time to learn languages is when the brain is receptive to these kinds of things, and that's much earlier, in preschool or elementary school."

Chugani believes that the new discoveries about the brain require a fundamental change in the nation's educational curricula. "There should be more emphasis on earlier education for key areas—language, music, math, problem-solving," he said.

Other countries have already done that, not with the insight that is coming from neuroscience, but from necessity. In Sweden, for example, which depends heavily on interactions with neighboring countries, children routinely learn three languages. Japanese children are subjected to intense education in their formative years so they have a chance to be selected for one of the limited spots at choice universities.

Documenting the times when the brain is biologically best equipped to learn is expected to have a major impact on society. "Understanding how the environment tunes up the brain during certain critical periods opens up a new frontier," said neuroscientist Tallal. "It gives you a powerful reason to say, 'Don't wait. You don't get another window of opportunity like that.'"

4 The Effect of Violence and Stress on Kids' Brains

A new kind of epidemic is ravaging our children, scientists warn. It is not caused by germs or poor diet, but by a scourge that is only now being recognized by medicine: brain damage caused by bad experiences.

Such damage, the evidence indicates, can increase the risk of developing a wide variety of ills ranging from aggression, language failure, depression and other mental disorders to asthma, epilepsy, high blood pressure, immune-system dysfunction and diabetes.

All of these problems are on the increase as the forces that generate stress—poverty, violence, sexual abuse, family breakup, neglect, drugs, lack of good stimulation, too much of the wrong kind of stimulation—continue to escalate. These kinds of bad experiences, pouring into the brain through the senses—sight, smell, taste, touch, sound—can organize the trillions of constantly active connections between brain cells into diseased networks.

"That puts a lot of importance on parenting because that has a big impact on the way the brain becomes wired," said Christopher Coe, a University of Wisconsin psychologist who has shown that infant monkeys deprived of parenting have deficiencies in key brain structures and suffer from numerous immunological disorders.

"There is a social cost if you don't have good parenting," Coe added. "It may be that you stamp an individual for their lifetime, not only in terms of their behavior and emotions, but literally their predisposition for disease."

One of the more astounding discoveries is that the stresses caused by bad experiences can actually affect genes, switching them on or off at the wrong times, forcing them to build abnormal networks of brain-cell connections.

"This means that the environment—external influences from conception onward—has a major role in shaping our individuality by shaping the expression of genes," said neuroscientist Bruce McEwen of Rockefeller University.

Bad experiences affect the brain primarily through the stress hormones such as cortisol and adrenaline. Designed to respond to psychological or physical danger, these hormones prepare the body for fight or flight. Normally such changes are smooth: The brain and body are prepared for action when need be and then put back on an even keel when the danger is over.

But when these hormones are overactive as a result of persistent stresses encountered during fetal development or early childhood, they can take over genetic regulation like a band of terrorists. The terrorized genes then set up aberrant networks of connections between brain cells, imprinting how the brain has mislearned: an epileptic seizure instead of a clear signal between cells, a depressive episode instead of a happy thought, a surge of rage instead of a willingness to compromise.

"We can now see how a learning disability could arise from a child's bad experiences," said neuroscientist Michael Merzenich of the University of California at San Francisco.

"It's not just by being born with a bad gene or a brain defect but by having a bad learning strategy from infancy," he continued. "We can see how the brain can become unstable and why that instability should result in a variety of neurological conditions that are common in humans."

How these bad experiences produce their damaging effects is only now beginning to be understood, and it is not without controversy, especially when genes are involved.

"Many people don't want to hear that your brain may be biologically different if you grow up in one environment or another," said Dr. Saul Schanberg, a Duke University biological psychiatrist. "One of those differences may be that [a stressful] environment has caused genes important for survival . . . to become overexpressed, making you more aggressive and violent," he added.

The brain is very resilient and maintains an even course in the face of the most outrageous experiences. That's why most children born in conditions of poverty and violence come out okay. Scientists suspect that the reason some children, regardless of their social or economic status, come out with damaged brains may be that they are genetically more vulnerable to stress. Furthermore, their bad experiences are not neutralized by a caring parent or involved adult.

"The things that are associated with resiliency have to do with protective factors like the quality of home life, the parent-child relationship, or another relationship that provides some security for the child," said Megan Gunnar, a child development psychologist at the University of Minnesota.

Animal experiments clearly show the protective power of a little security, and the brain damage that can occur when it is absent. Newborn animals that are deprived of nurturing by their mothers become dysfunctional and antisocial.

Such damage also occurs before birth, a period in which the fetus was once thought to be protected. For example, rats that are stressed during pregnancy give birth to offspring that are very emotional and reactive.

"They have normal offspring from the standpoint of size and appearance," said Dr. Ned Kalin, chief of psychiatry at the University of Wisconsin at Madison. "But when you look at their development you find that they are hyperresponsive to stress. When you look at their brains you find more adrenaline [a stress hormone]."

On the other hand, Rockefeller University's McEwen and others have found that when they can, in effect, turn up the volume on mothering, newborns grow up in the opposite way: calm, cool, and ready to explore. In their experiments, rat pups are removed from their cages for fifteen minutes a day and then immediately returned.

The difference is that the worried mothers showered attention on the handled rat pups after they were returned to the cage, thereby turning down the amount of stress hormones their young brains would otherwise have been making, McEwen said. "People need to be aware that the brain is doggone vulnerable," he said. "If something happens early in life it can have permanent consequences for how a kid develops and learns."

For an increasing number of children, bad experiences are on the rise. The Census Bureau reports that in 1991 there were 14.3 million children living in poverty, an increase of 1 million from 1990.

Violence has become an overwhelming and mind-shattering way of life for many youngsters. A study of more than 1,000 students from poor Chicago neighborhoods found that 74 percent of them had witnessed a murder, shooting, stabbing, or robbery. Nearly half of them were themselves victims of a rape, shooting, stabbing, robbery, or some other violent act.

"The brain responds to experience," said Dr. Richard Davidson, a professor of psychology and psychiatry at the University of Wisconsin. "Children who are raised in impoverished conditions . . . are at high risk for having impoverished brains."

The magnitude of the problem was revealed in a recent nationwide study showing that one child in five under age eighteen has a learning, emotional, behavioral, or developmental problem that researchers say can be traced to the continuing dissolution of the two-parent family.

Forty-two percent of U.S. families with children start out with one, two, or three strikes against them, said psychologist Nicholas Zill of Child Trends, Inc., a Washington organization that studies social changes affecting children.

The first strike is lack of education: The mother has not finished high school by the time she has her first baby. The second strike is lack of commitment: The mother and father are unmarried when they have their first child together. The third strike is lack of maturity: The woman is under twenty when she gives birth for the first time. One new family in nine has all three strikes against it.

All these statistics point to the fact that many young parents today may be less prepared to care for children than were their predecessors.

"The biology of our species makes necessary a huge parental investment in order to achieve the fulfillment of each child's potential," said David A. Hamburg, president of the Carnegie Corporation of New York. "For all the atrocities now being committed on

NEGATIVE EXPERIENCE AND THE BRAIN

Brain connections can be damaged by negative experiences, making learning more difficult and sometimes resulting in violent behavior.

MENTAL DISORDERS AND CRIME

A study done by scientists at the University of Montreal shows that substance abusers and subjects with major mental disorders were more likely to be convicted of a crime than those with no disorder or handicap.

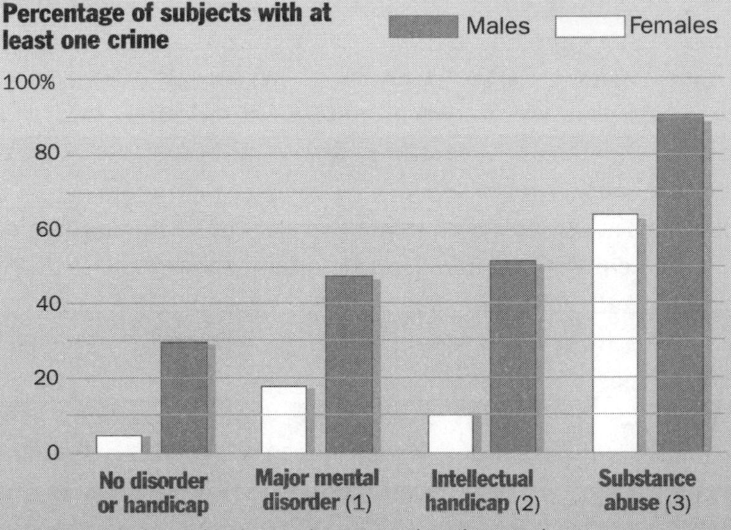

1. Schizophrenia, major depression, paranoia, other psychoses.
2. Attended special classes for intellectual deficiency but were never admitted to a psychiatric ward.
3. Alcohol and/or drug abuse.

"Developmental experiences determine the capability of the brain to do things. If you don't change those developmental experiences, you're not going to change the hardware of the brain and we'll end up building more prisons."

—Bruce Perry, neuropsychiatrist

Sources: Chicago Tribune, University of Montreal.

Chicago Tribune

our children, we are already paying a great deal . . . in economic inefficiency, loss of productivity, lack of skill, high health care costs, growing prison costs, and a badly ripped social fabric," he said.

New research is redefining the roles of nature and nurture in determining how a child will turn out. In the past, scientists argued that one or the other was more important, but the contemporary view is that both are constantly in play. Such information is beginning to better define the conditions that put children at risk for disease. Finding these children and altering the detrimental experiences as early as possible can change the course of their lives.

Jerome Kagan, a Harvard University psychologist, has been trying to do just that. His studies of middle-class Boston children revealed that about one in three had psychological problems primarily related to a bad environment.

"The causes are always in the biology of the child, either a certain neurochemistry you inherited, structural abnormalities that occurred prenatally, or a bad environment," he said. "And a bad environment—strife at home, abuse, bad peers, lack of role models—is always the most prevalent cause."

Discoveries in the last five years have revolutionized how scientists think about the impact of such negative experiences on brain development. One of the most profound is the finding that environmental stress can activate genes linked to depression and other mental problems.

Research led by Robert Post, chief of the National Institute of Mental Health's biological psychiatry branch, found that stress or drugs of abuse, like cocaine and alcohol, can turn on a gene called C-fos.

The protein made by the C-fos gene attaches to a brain cell's DNA, turning on other genes that make receptors or more connections to other cells. (Receptors are little doorways that sit on a cell's surface to let hormones and other chemical messengers in or out.) The problem is that these new connections and receptors are abnormal. They cause a short circuit in the brain's communication networks that can give rise to seizures, depression, manic-depressive episodes, and a host of mental problems, Post said.

Stress, for instance, through its hormonal intermediaries, turns on genes that leave a memory trace of a bad feeling. Then along comes a lesser stress that triggers the same memory trace and reinforces it. Now, instead of a lousy feeling, the person gets depressed. Finally, after repeated reinforcements, the memory trace takes on a life of its own, firing willy-nilly and producing depression without any outside trigger, Post said.

"The idea that you can learn bad things like depression and epilepsy and that they are encoded through the genes into the physical structure of brain cells is new and exciting," he added. "It provides some of the molecular mechanisms to explain what scientists are beginning to suspect and fear can happen to people who have horrendous developmental experiences."

Such findings already have paid big dividends. Anticonvulsant drugs used to stop epileptic seizures are now routinely used in manic-depressive patients to block the short circuits that precipitate their highs and lows.

Martha Pierson, a neurobiologist at the Baylor College of Medicine in Houston, demonstrated how she has discovered that bad developmental experiences can produce deadly seizures 100 percent of the time in laboratory animals.

She found that if newborn rats, which start hearing on day thirteen, are prevented from hearing sounds for the first two days of this critical period, the connections among their brain cells do not organize into normal patterns. When an animal finally hears a normal sound, it gets an immense input of signals, like being in a garbage can with someone banging on the lid.

"Their brain wiring is scrambled," Pierson said. "They attack anything in sight for five to ten seconds and then go into convulsions. They would die in another seven seconds if you didn't revive them with cardiac massage."

Clearly, the lack of sound during a critical period in development can cause one type of epilepsy. "It means that with bad experiences, you can fail to learn [and] you can develop a disease," she added.

The new findings are helping to explain the big increase in mental problems among America's youth, and to refocus the goals of the National Institute of Mental Health.

"How the brain interacts with the environment, especially during the critical periods of development, has become central to our mission because most mental disorders have their onset in childhood or adolescence," said the institute's former director, Dr. Frederick Goodwin.

An epidemic of mental problems has caught researchers off guard. In the last twenty-five years, Goodwin points out, there has been a doubling of the rates of depression, suicide, crimes of violence, drug abuse, and alcohol abuse.

This dramatic increase comes at a time when the world of many children is unraveling—the divorce rate is doubling, parenting time is reduced because both parents are more likely to be working, the mobility of the population has increased, poverty is growing.

"These trends may be more exaggerated in the inner city but, in fact, they are across the board," Goodwin said. "You only have to look into some of our suburban neighborhoods to see the same problems. It's just that there are more resources available in the suburbs to help these kids."

Understanding how the brain can be damaged as a result of bad experiences gives scientists a new opportunity to prevent the damage and to repair it once it has occurred.

"The question is not only, 'What's wrong with the environment and what can we do about it?' but, 'What makes some kids more vulnerable than others and how can we develop ways to protect them?'" Goodwin said. "That's the new direction we have to go in. If we did that, we'd need fewer prisons."

At the University of Chicago, scientists are tracking down the environmental inputs that may direct a brain down a path of aggression and violence.

The main culprit is stress. Many children are raised in violent, abusive surroundings over which they have no control, said neuropsychiatrist Dr. Bruce Perry, a leader in the work at the University of Chicago, who has since moved to the Baylor College of Medicine. The antidote is giving children a sense of self-worth and teaching them that they are not helpless. "If there's somebody out there who makes you feel like you're special and important, then you can internalize that when you're developing your view of the world," Perry continued. "When you look at children who come out of terrible environments and do well, you find that someone in their lives somehow instilled in them the attitude that they aren't helpless, that they aren't powerless, that they can do something."

Many never get that antidote in time. They often are verbally impoverished on the one hand, but are extremely rich in stressful experiences on the other. They are above average in reading nonverbal cues that tell them when others may be threatening or vulnerable—a capacity that has come to be called "street smart."

But they get into trouble by misinterpreting some of these visual cues, like the student who throws a tantrum or drops out of school because he views a teacher's criticism as aimed at his self-worth rather than as an effort to help him learn. "Their brains are different because of bad environmental experiences," Perry said. "That makes them at risk for the development of a variety of cognitive, behavioral, and emotional problems, and puts them at greater risk for developing certain neuropsychiatric disorders."

Stress hormones rachet up all of their reactions so that their hearts beat faster, their blood pressure is higher, and they are more impulsive than a normal youngster. And these findings may explain the high rate of hypertension in black males, Perry said. Medical researchers used to think that it was genetic, but now it appears to be caused by stress. "These kids are doubly at risk," explained Perry. "They don't have the opportunities to learn the traditional currency by which we normally get along in our society, and their brain systems that are involved in mood and impulsivity are poorly regulated.

"As they get older, these kids have fewer coping skills and fewer ways to solve problems. That predisposes them to use aggressive and violent strategies to try to solve problems."

But just as anticonvulsant drugs can control manic-depressive episodes by quelling short circuits in the brain, Perry has found that an antihypertensive drug can reduce aggressive tendencies in these supercharged youngsters as well as lower their blood pressure.

Preliminary results indicate that the drug, called clonidine, blocks much of the action of adrenaline. Adrenaline is a major stress hormone that increases blood pressure, speeds the heartbeat, tightens muscles, and in other ways prepares the body for action in emergencies. When the emergency is over, adrenaline normally recedes and the body calms down. In these children, however, the hormone is kept on high alert.

"They are hyperaroused, impulsive, and have difficulty concentrating," Perry said. "Each of the kids we've tested clonidine on has had a significant reduction in symptoms."

Studies show that every dollar spent on early childhood development programs translates into saving five dollars later in social services, mental health services, prisons, and other programs intended to deal with the aftermath of aggression and violence, he said. "Developmental experiences determine the capability of the brain to do things. If you don't change those developmental experiences, you're not going to change the hardware of the brain and we'll end up building more prisons."

5 Averting Retardation through IQ Boosts

I am not the man I was.
—Ebenezer Scrooge, *A Christmas Carol*

Pushing the right biological buttons in the brain, scientists are finding they can make the future brighter for many children whose development otherwise would have been stunted. Based on preliminary results, they can boost IQ levels by ten to twenty points, reduce mental retardation by 50 percent, and cut school failure rates by much more.

That there are biological buttons, and that they work to physically reshape the brain by engaging its genetic gears and hormonal levers, is only now being understood. How the buttons work is perhaps the most amazing thing of all. The buttons are the senses—vision, taste, smell, touch, sound—and they can be pushed by experiences from the outside world.

The environment plays on the brain like a computer keyboard, inputting good experiences as well as bad ones, and, often, no experience at all. All of this is dutifully laid down in the trillions of connections between brain cells that make learning and memory possible. What happens to children's brains when the buttons are pushed, or not pushed, can determine their IQ and whether they will become mentally retarded, sick, aggressive, violent, and even if they will live or die.

Failure to push certain buttons creates abnormal brain circuits that can now be seen as the cause of a vacant stare or a menacing grimace.

With new high-tech imaging devices that can peer into the living brain, scientists are beginning to see what is going wrong, and they are figuring out which buttons haven't been pushed—and need to be—to make the brain work right.

“We’re going to be able to identify these [brain-impaired] children within the first five years of life and make interventions to change their brains right then and there,” said Dr. Ned Kalin, chief of psychiatry at the University of Wisconsin at Madison.

Children someday may be immunized against stunted brains. Such immunization may be in the form of environmental interventions, such as new learning experiences, or perhaps a combination of environmental changes and drugs. “Just as an enriched environment can change the chemistry of the brain for the better, exposure to the right drug at the right time may reset the development of the brain in a positive way,” Kalin said.

Harry Chugani, a pediatric neurologist at Wayne State University in Detroit, agrees. Chugani, whose imaging studies revealed that children’s brains learned the easiest and fastest between the ages of four and ten, said these years are often wasted because of a lack of input.

“We can head off a lot of the problems we end up having to deal with down the road by beginning early to teach kids all kinds of things, such as a second language, math, and musical instruments,” he said. It is modern science’s version of Pygmalion. “You can completely change the way a person will turn out when you do that,” said Chugani.

So new is the evidence linking early experience to major biological changes in the brain that it has been only in the last ten years that the scientific community has come to accept the fact that it really happens.

“It used to be that we thought the brain was hard-wired and that it didn’t change,” said Dr. Frederick Goodwin, former director of the National Institute of Mental Health. “What you were born with, like your IQ, was fixed. Now, of course, what is being learned is, ‘Hey, these environments can actually produce physical changes in the developing brain.’ ”

Not only that, but the new imaging technology will be able to measure those physical changes to determine how well the intervention programs are working.

One of the first scientists to discover what happens when the right biological buttons are pushed is William T. Greenough, a pioneering psychologist and cell biologist at the University of Illinois at Champaign-Urbana. When Greenough exposed rats to an enriched environment full of toys, food, exercise devices, and playmates, he found, on autopsy, that they had super brains.

The brains of the experience-enriched rats had about 25 percent more connections between brain cells than those of rats raised in standard, drab laboratory cages. Tests showed that the enriched rats were a lot smarter than the deprived ones.

Craig Ramey of the University of Alabama found that he could do the same thing with children and produce similar results.

Starting with children as young as six weeks, he exposed a group of impoverished inner-city children to an enriched environment—learning, good nutrition, toys, playmates. A similar group of children were used as controls.

Ramey tested their IQs after twelve years of age, and the benefits of early intervention endured: The enriched youngsters had significantly higher IQs than the control group. PET scans, which measure brain activity, showed that the brains of the children exposed to stimulating environments were perking along at a more efficient rate than those of the control children.

"This gives us reason to believe that the positive changes we are seeing on the behavioral level are, in fact, undergirded by an increase in synaptic connections between brain cells," Ramey said.

Remarkably, the enriching experiences also prevented mental retardation. Children in the control group, whose environment remained impoverished, had a high rate of preventable retardation.

"The bottom line is that we now have clear and strong evidence that if we begin early intervention, in the first year of life, we can prevent a very substantial amount of mental retardation and developmental disabilities," said Ramey.

What Greenough, Ramey, and others are showing is the biological basis for early learning—Greenough showing why it is true, Ramey showing that it is true for humans as well as animals.

"Our work gives a kind of biological reality to early childhood education," Greenough said. "The brain really is sensitive to the early environment, and those early experiences are likely to underlie the kinds of intellectual performance that we see on the behavioral side in humans."

The new findings are expected to have a revolutionary impact on early child-rearing practices and education, and provide further support for such intervention programs as Head Start. The findings also lend support to the view that society cannot leave children alone to fend for themselves in impoverished environments that dull their brains or hostile environments that twist their brains into nerve centers for violence.

The quality of the environment and the kind of experiences children have may affect brain structure and functioning so profoundly that they may not be correctable after age five, Ramey said.

"If we had a comparable level of knowledge with respect to a particular form of cancer, or hypertension, or some other illness that affected adults, you can be sure we would be acting with great vigor," he added. "The problem is that young children don't vote, and they can be sort of kept hidden for a while until they show up in the school system failing miserably," said Ramey. "We are creating a backlog of children who have a great criminal potential."

Ramey's intervention program includes good health care, not only for the child but also for the mother, and it helps parents or caregivers to improve their own lives through education or employment opportunities. But the greatest emphasis is placed on the children who spend five days a week in a special educational program until age five.

The first clear evidence that Ramey's interventions were working came after three years. Children in the intervention group had IQ scores in the normal range, around 100, whereas the average IQ score of children in the control group was 20 points lower.

Early enriching experiences also were able to thwart the mind-stunting effects of little or no stimulation. All of the children in the intervention group whose mothers had IQs below 70 climbed the intellectual ladder as a result of the stimulating experiences and had normal IQs.

The effect of not intervening became tragically evident in the group of control children. The children of low-IQ mothers, who were mired in the same intellectually impoverished environment as those mothers, likewise became borderline mentally retarded.

"Early intervention appears to have had a particularly powerful preventive effect on children whose mothers had low IQs—while also benefiting other children from economically, socially, or educationally disadvantaged backgrounds," Ramey said.

The benefits of intervention rolled into the school years. While 50 percent of the children who did not receive intervention failed one or more grades by the age of twelve, only 13 percent of the intervention children had similar failures.

The annual cost of such early intervention is between seven and ten thousand dollars for each child. But cost-benefit analysis shows that by preventing mental retardation and school failure, such programs return an estimated three dollars for every dollar spent on them, said Ramey. And if follow-up studies determine that intervention reduces crime and unwanted pregnancies, then the savings to society from such programs would be much higher.

Seventy-five percent of all imprisoned males in America have poor school records and low IQs, Ramey pointed out. Tracing their backgrounds turns up a familiar pattern: They begin as children from disadvantaged families starting school academically behind. They don't know how to read or do basic math; because they are in poor school systems, they get little help. Their growing frustration often turns into truancy, school failure, aggression, and violence.

"There's no real mystery about this. When you have high concentrations of people who don't have basic social skills—and being able to succeed in school is a universally required basic social skill—you have chaos," Ramey said.

The lack of stimulating experiences and intellectual enrichment is a problem not only in poor neighborhoods. It has spread through middle class and wealthy populations, as the dramatic upheavals in the traditional family cast an increasing number of youngsters adrift.

Unlike many European and Asian societies, which invest time and energy in their children, American society is becoming less likely to do that.

The enemy is "time poverty," said Felton Earls, professor of human behavior and development at the Harvard School of Public Health, and professor of child psychiatry at the Harvard Medical School. Many parents, regardless of their income, do not have enough time to organize a stimulating environment for their children.

Not only are more women working than ever before, but men are working longer hours. Children are often left to fend for themselves, surrendering to the passive habit of watching TV, instead of interacting with their environment.

"We have a situation in which parents in all social classes have less time to put into the management and orchestration of what a child needs during this critical period of time," Earls said. Day care must be upgraded nationally to provide the best kind of learning environment for children when their parents are unable to do so.

"Good day care can solve a lot of the problems because children interact a lot with other children," said Martha Constantine-Paton, a Yale University neurobiologist. "People are designed to interact with other people."

Dr. Norman Krasnegor of the National Institute of Child Health and Human Development says there is a special kind of "dance" between a mother and her baby that is missing in many families. "What we have learned is that human babies elicit behavior from their parents and the parents, in turn, give of themselves," he said. "It's not a one-way street. It's a kind of a dance between mother and child."

Many parents do that intuitively, using mobiles, talk, and other devices or activities to stimulate reactions from their babies. But if the mother is a teenager, lives in poverty, or is on drugs, then she may not be responsive to what the baby is trying to elicit, and the child's brain may not receive the stimulation it needs to develop.

Krasnegor called for the establishment of visiting nurse programs to help educate mothers about child care and to let them know that their infants need to receive appropriate stimulation. Such stimulation may mean the difference between life and death for very premature infants.

Working first with animals and then with human "preemies," Dr. Saul Schanberg of Duke University and Tiffany Field of the University of Miami discovered one of the most important biological buttons of all: touch.

Isolated in incubators bearing Do Not Touch signs, preemies struggled to survive. Since they were so tiny, doctors felt that they should not be disturbed. Anything that caused them to cry endangered their breathing.

But no matter how well the preemies were fed and their medical needs tended to, most of them didn't grow. They were stuck on pause and many became physically and mentally retarded, or didn't make it at all. In a series of experiments, Schanberg and Field found out why.

Separating newborn rats from their mothers caused the pups to go into a survival mode. To conserve food and energy, their bodies stopped growing. Stress hormones, released to subdue the body's need for nourishment, actually turned off genetic activity so that cells could not divide.

When the mothers were returned, the stress hormones in their pups subsided and they began growing again. But when the scientists anesthetized the mothers before returning them, the pups failed to revive.

Eliminating one possible cause after another, the scientists found that it was the mother's licking that kept their newborns happy and the stress hormones in check. Licking was the signal that told the pups that they were not in danger. Simply swabbing abandoned pups with a wet paint brush could do the same thing, because it mimicked the mothers' licking and allowed the newborns to thrive.

When the researchers looked at preemies, sure enough, the same chemical changes were happening. Cortisol, a major stress hormone, was up and DNA synthesis was down.

Human babies are not licked, but they are held and their backs are rubbed. When Schanberg and Field tried that on the premature babies, the infants started to grow stronger and thrive.

Before the touching therapy started, the preemies were growing at the anemic rate of twelve to seventeen grams a day, less than half the rate of growth they would have been experiencing inside the womb.

Touching and rubbing shot their daily growth rates up to thirty grams, about an ounce. They were able to leave the hospital six to seven days earlier than nontouched preemies, at a savings of $5,000 per baby.

As with the rats, the lack of touching serves as a signal that a newborn's mother and food supply are not available, thereby kicking in the cortisol-driven survival mechanism. "Once we established that touching was not just a fairy tale, it became medically acceptable. This wasn't just being nice, this was, 'You better damn well have it or else,'" said Schanberg.

Good news like that spread rapidly. Most hospitals now have a program involving grandmothers and others who regularly touch and massage extremely low-weight infants.

Threats to brain development can come from unexpected sources. One, which affects millions of American children, is the newly recognized danger from undernutrition. Unlike severely malnourished children, who obviously appear stunted, both mentally and physically, youngsters who are undernourished may appear normal, and the peril to their intellectual development is more subtle. In the United States, mild to moderate undernutrition is a major problem, affecting an estimated twelve million children.

"In general new research findings show that lack of sufficient food during childhood, even on a relatively mild basis, is far more serious than previously thought. By robbing children of their natural human potential, undernutrition results in lost knowledge, brainpower and productivity for the nation," concluded a Tufts University School of Nutrition report.

Undernourished children usually are fatigued and uninterested in their social environment. Limited food energy is conserved first for maintenance of vital organ functions, then for growth, and finally for social activity and cognitive development. Children who are chronically underfed become less active and more apathetic, a condition that cheats them of normal social interactions, inquisitiveness, and intellectual functioning. The ranks of school dropouts are filled with undernourished children.

Fortunately, undernutrition's threat to mental development can be easily reversed and prevented, the Tufts report points out. Nutrition and prenatal care for women reduce the incidence of low-birth-weight babies and subsequent developmental delays. Treating iron deficiency, which affects one out of four poor American children, can prevent the effects of anemia on learning, attention, and memory.

And research consistently establishes that federal initiatives such as the School Breakfast Program and the Special Supplemental Food

Program for Women, Infants and Children (WIC) have positive effects on the cognitive development of children. The benefits include higher performance on standardized tests, better school attendance, lowered incidence of anemia, and reduced need for costly special education.

At the University of Wisconsin, Dr. Richard Davidson, professor of psychology and psychiatry, is using brain-imaging devices to look for particular patterns of brain organization that may predispose a child to be happy or fearful. So far, he can accurately predict whether a ten-month-old infant will cry or not in response to a brief period of separation from his mother.

Those who cry have less electrical activity in the left-front section of their brains than those who don't cry. They also are more reticent and less willing to explore than infants who have a higher level of electrical activity in the left sides of their brains.

"These individual differences are stable, trait-like attributes," Davidson said. "They show up early in life and they are consistent over time." Davidson hopes that by making pictures of these brain traits it may be possible to detect children who are prone to developing mental disorders, such as depression, long before the symptoms show up.

But what really fascinates him is the possibility that these negative patterns can be changed through new learning experiences. The wiring that gets established in the brain and makes a child fearful and hesitant may be undone, then formed again into networks that make the child more outgoing and trusting.

Youngsters who have a lower level of left-frontal brain activity are more likely to believe that when bad things happen to them it's their own fault. Children with higher left-brain activity, on the other hand, tend to look outside of themselves for the cause, and then try to correct it.

"It may be that training a person how to think differently about the causes of negative life events will actually reduce the chance of their developing some psychopathology by changing the pattern of brain activity it's associated with," Davidson said.

Part Two HOW THE BRAIN GETS DAMAGED

6 Tracking Down the Monster within Us

Scientists know that the devil does not make people commit murder, rape, assaults, suicides, or other acts of violence. But until just recently they didn't have a better candidate. All they had were puzzles. Why do some children grow up to be violent criminals while their brothers and sisters turn out to be law-abiding citizens? Why do some people who would never hurt a fly, suddenly break and turn violent?

For the first time scientists are beginning to find answers to these questions from an explosion of new information. In a startling series of recent breakthroughs, researchers probing the biology of violence have pinpointed how aggression is triggered in the brain—and how it might be prevented.

Underlying the scientific quest, which has revealed genetic and environmental links to abnormal brain chemistry, is the growing suspicion that society may unwittingly be feeding the nation's growing epidemic of murder, rape, and other criminal acts by making childhood more dangerous than ever.

The most profound discovery is that genetic defects produce abnormal levels of serotonin and noradrenaline, two potent brain chemicals that researchers have successfully manipulated to make animals more violent or less violent. Several studies also suggest that threatening environments can trigger serotonin and noradrenaline imbalances in genetically susceptible people, laying the biochemical foundation for a lifetime of violent behavior.

Such ominous trends as the collapse of the family structure, the surge in single parenting, persistent poverty, and chronic drug abuse can actually tip brain chemistry into an aggressive mode—an effect that was once thought impossible.

These and other findings raise the hope that violent behavior eventually can be curbed by manipulating the chemical and genetic keys to aggression. The findings also raise fears.

The link between brain chemistry and violence "is extremely interesting and fascinating," said Marie Asberg, chief of neuroscience at the Karolinska Hospital in Stockholm. "But it's also a bit dangerous.

"It's the power in it. If we know the biology of how aggression works, we might also be able to manipulate it in ways that we don't like. We don't want to stigmatize people as aggressive . . . because of their brain chemistry. But we do want to use it to help them."

The prospect of being able to determine a person's propensity for violence by measuring levels of brain chemicals—and then regulating those levels—raises several ethical issues. It is virtually certain, for example, that simple screening tests will be developed to determine levels of serotonin and noradrenaline. Antiviolence medications conceivably could be given, perhaps forcibly, to people with abnormal levels.

There also is concern that genetic screening would lead to preventive detention—the jailing of those found to be genetically predisposed to violence.

Some people reject any suggestion that genes influence behavior, arguing that people aren't slaves to their genes. This is especially true when genetics and violence are linked. Such an association raises fears that some groups will be stigmatized as violence-prone, even though researchers emphasize that genes linked to aggression are found in all racial and ethnic groups.

Most of those who study violent behavior believe the implications of their findings are liberating, not threatening. In demonstrating that self-control and impulsiveness may be regulated by brain chemistry and genes, their research challenges the concept that violent crime is the result of an evil will; rather, it places aggressive violence in the same category as depression, schizophrenia, and other mental disorders.

Researchers add that the findings do not negate the concept of free will. Their view is that genes make certain behaviors much more likely, but not inevitable.

The new insights into the biological roots of violence stem from a revolution in molecular biology, the attempt to understand life molecule by molecule.

As the chemical shadows of aggression flit through the corridors of the mind, scientists can spy on them with new imaging techniques that peer inside churning brains and decipher the molecules of thoughts and emotions. And using the same research tools that led to the discovery of the genetic causes of cancer and other diseases, scientists are venturing into the largely unexplored relationship between genes and behavior.

Aggression is not necessarily a bad thing, scientists point out. It evolved as a positive force in human evolution, enabling people to compete for food, mates, shelter, and status. Aggression is universal, used by all vertebrates for survival and reproductive advantages. "Clearly aggression at some level is important in allowing us to have the courage to live our lives and stand up for ourselves," said Yale psychiatrist Dr. John Krystal.

Normal aggression has a set point, like body temperature, which is regulated by brain chemicals. Most people are born with a balance of these chemicals that enables them to react to events in reasonable ways. But changing that set point can either increase aggression or lower it.

Researchers are learning how this set point can be altered, and they have found that the mechanism for change—an imbalance of serotonin and noradrenaline—is shared by humans and animals. The two chemicals are known as neurotransmitters, which enable brain cells to talk to each other in peace or squabble in anger.

Serotonin is the brain's master impulse modulator for all of our emotions and drives. It especially keeps aggression in line. When serotonin levels fall, violence rises, like some long-subdued monster breaking free of its bonds.

Noradrenaline is the alarm hormone. It organizes the brain to respond to danger, producing adrenaline and other chemicals that prepare the body to fight or flee. Noradrenaline may play a major role in both hot-blooded and cold-blooded violence. When noradrenaline is turned on "high" and left there, impulsive violencc of the hot-blooded type becomes more likely.

Low levels of noradrenaline, on the other hand, cause under-arousal. To get their thrills, many people with low noradrenaline take calculated risks, sometimes of the type associated with predatory violence—the premeditated or cold-blooded kind that may be found in a serial killer.

Serotonin and noradrenaline may work separately or together in different combinations to produce a spectrum of violent activity. So basic is their teeter-totter relationship that serotonin increases during sleep and decreases during wakefulness, while noradrenaline increases during wakefulness and declines during sleep.

At normal levels, serotonin keeps in check primitive drives and emotions that have taken millions of years of evolution to subdue—sex, mood, appetite, sleep, arousal, pain, aggression, and suicidal behavior.

Such control is exerted through the neocortex, the part of the brain that oversees socialization, memory, and judgment and sits like a convoluted crown controlling the deeper parts of the brain that harbor primitive instincts and emotions.

Anything that happens to the thinking, outer surfaces of the brain, such as injury or bad learning experiences, can interfere with serotonin's ability to keep the mob of basic instincts in line.

When serotonin declines, for instance, as it does in the brains of abused children or in people using alcohol and possibly cocaine, steroids, and other substances, impulsive aggression is unleashed.

One reason that aggressive youths tend to mellow as they get older, researchers suspect, is that serotonin levels increase with age. And the reason that females generally are less aggressive than males may be because their serotonin levels are 20 to 30 percent higher.

A low serotonin level also can dry up the wellsprings of life's happiness, withering a person's interest in his existence and increasing the risk of depression and suicide. Alcoholism, sleeplessness, sexual deviance, fire-setting, obesity, and other impulse-control disorders also have been laid at the doorstep of low serotonin.

A growing body of evidence indicates that low levels of serotonin are implicated in a lack of control, the kind of behavior that typically manifests itself as irritability, loss of temper, and explosive rage. It is the type of impulsive aggression that is escalating at an unprecedented pace in the United States.

While the U.S. population increased by 40 percent from 1960 through 1991, violent crime increased 560 percent, murders increased 170 percent, rapes 520 percent, and aggravated assaults 600 percent, according to the FBI.

"Violence in America is clearly a public health, public safety, and mental health disaster," said psychologist Dean G. Kilpatrick, of the Crime Victims Research and Treatment Center of the Medical University of South Carolina in Charleston.

But the dramatic statistics tell nothing of what is going on inside the brain to cause the eruption of violence. Until now, scientists have been stymied.

The best they could come up with was a profile: Childhood aggression leads to adult violence; those most susceptible to violent behavior have low IQs, poor school attainment, high impulsivity, and lack of concentration; they tend to come from big families, low-income families, and have parents who were convicted of a crime; they grew up with harsh discipline, poor supervision, and separations from parents.

But most people exposed to these conditions manage to avoid taking the path to violence. Why do they escape while others, who run the same risky physical and psychological gamut, don't?

The answer appears to lie in their genetic differences. Two gene mutations recently have been found that neatly support the growing evidence of the genetic-environmental link to violence.

Researchers from the Netherlands, led by Hans G. Brunner of the University Hospital in Nijmegen, stunned the scientific community with their discovery of a rare mutant gene that raises noradrenaline levels and increases impulsive aggression in the men who possess it. The discovery, reported in the journal *Science*, was believed to be the first that showed a gene defect can change behavior.

But scientists at the National Institute on Alcohol Abuse and Alcoholism, headed by Markku Linnoila, have made an even more far-reaching discovery—a mutant gene that appears to be widespread in the populations so far studied that lowers serotonin levels and increases aggression.

The quest for the biological roots of violence—and there may be others besides serotonin and noradrenaline—draws on a wide variety of research, including studies of insects, monkeys, reproduction, and heart disease, in addition to brain chemistry.

None provides a more vivid example of the environmental-genetic link to violent aggression than the Grand Canyon Tiger Salamander, nature's version of Dr. Jekyll and Mr. Hyde.

The salamanders live in ponds along an isolated rim of the Grand Canyon. When water and food are plentiful, the salamander is in its Dr. Jekyll form—a gregarious, peace-loving insect eater. But when the water begins to dry up, food becomes scarce and living conditions become unbearably cramped. Then some of the salamanders go through an amazing Mr. Hyde transformation.

Environmental pressures rapidly alter the function of some of their genes, creating changes in their physical shape and making them aggressive. Muscles enlarge to make their heads and mouths bigger and they grow a new set of huge teeth, adaptations that allow them to attack and eat other salamanders.

They become cannibals, but only for a short time. Once they've gobbled up enough salamanders to reduce crowding, they turn back into Dr. Jekylls. Their heads shrink to normal size, their cannibal teeth disappear, and they dine on insects again.

David Pfennig, a Cornell University behavioral ecologist who is studying the Grand Canyon Tiger Salamander, said many other species undergo cannibalistic transformations as a result of environmental pressures such as overcrowding.

Similarly, research scientists can make a monkey so violent that it is ostracized from a community, simply by changing the way it is brought up or by altering its genes.

Researchers usually confine their studies to animals such as monkeys, which share 95 percent of human genes, and salamanders because of ethical limitations on how much research they can do on people. But people can experiment on themselves. And they do, in effect, with their child-rearing practices and lifestyles.

These inadvertent experiments are being carefully watched because they may be producing the kinds of aggression and violence in humans that scientists have provoked in laboratory animals.

At the top of the list of these "natural experiments" is the growing number of babies born to unmarried teenagers living in overcrowded and poor environments. They are at the forefront of a tidal wave of unmarried motherhood, and they are the most vulnerable part of that trend.

In 1990, teenage mothers accounted for 1 of every 7.7 births and 68 percent of these adolescents were unmarried, an increase of 127 percent since 1970, according to the National Center for Health Statistics. In Chicago, 89 percent of teen mothers were unmarried.

When mothers themselves are still children, their babies often are deprived of the positive experiences that establish appropriate learning circuits in their brains. At the same time, they often are the victims of violence or witness it.

Sensory deprivation and exposure to violence are detrimental early experiences that are known to cause physical stunting of the brain.

Researchers fear such experiences may be responsible for twisting the architecture of the brain into abnormal networks. Crippled networks may push noradrenaline production into overdrive or serotonin into low gear, establishing impulsive thought processes that view violence as a way of dealing with problems.

The statistics supporting this link are chilling: Most state prison inmates in the United States did not live with both parents while growing up and the majority were born to teenage mothers, the U.S. Bureau of Justice reports. As of 1991, these inmates had 826,000 children under the age of eighteen.

According to 1991 statistics, one of four American children lives with a single parent and one in five lives in poverty.

When these statistics are compared to those of other developed nations, American children are worse off, even when minorities and the poor are excluded, said Cornell University developmental psychologist Urie Bronfenbrenner, a founder of the national Head Start program.

"There are grounds for believing that families and children are becoming an endangered species, and perhaps dangerous as well," he said.

Violence may also arise from other sources, many of them unexpected.

- Some people experiment on themselves with steroids. More than a million Americans, primarily men, use synthetic sex hormones to build muscles. But some get more than they bargain for—an imbalance in brain chemistry that leads to aggressive and violent behavior.

- Alcohol has long been known to provoke aggression and violence in some people. But only now are scientists beginning to learn some of the ways that alcohol can alter brain chemistry by lowering serotonin.

- Many other people may be dieting themselves into aggression. Preliminary findings indicate that low-cholesterol diets, which are followed by millions of heart-conscious people, may provoke impulsive or aggressive behavior.

- Still another form of self-experimentation with violence involves cocaine. Studies of people who take cocaine show that the mind-altering drug can affect levels of serotonin and other important brain chemicals. Many scientists believe that this may account for the violence seen in some cocaine users.

- Cocaine, alcohol, and other drugs of abuse, which are taken by an increasing number of pregnant American women, may also reset the chemical balances in the brains of their fetuses. Drug use in pregnancy already is linked to an increased rate of hyperactivity in children.

- Brain injury is a well recognized cause of violence. Of the 70,000 to 90,000 Americans who annually suffer brain damage resulting from injuries, many will experience altered brain chemistry and engage in explosive violence. Most Death Row inmates who have been studied, for instance, are victims of brain injuries.

◦ Exposure to lead from leaded fuels and paint and plumbing in older houses causes learning problems and hyperactivity in many children, often the first steps toward violent behavior.

New insights about the causes of violent behavior are also challenging the notion that violence is something a person always does on purpose and should be punished for. Now it appears that many types of violence, especially impulsive violence, may be just like other mental disorders—a dysfunction of the brain—that can be treated and prevented.

"We view people who are violent in the same way we used to view people who were mentally ill. In the old days, schizophrenics, manic-depressives, and others were thought to be bad people who had to be punished," said Dr. Stuart Yudofsky of the Baylor College of Medicine in Houston.

It was only in the late 1950s and early 1960s, with the discovery of the first antipsychotic drugs, that science confirmed the biological basis of mental illness and developed successful new treatments for it. Until then, the increasing number of mental patients could be accommodated only by building more mental hospitals.

That building spree came to an abrupt stop with the advent of psychoactive drugs, and most mental hospitals have since been emptied of their formerly untreatable patients. In Illinois, for instance, the number of mental hospital beds shrank from a high of 54,000 in 1958, to about 3,200 today.

"When we reconceptualize violence as involving the brain, then we are really going to start making progress," said Yudofsky, who developed the first drug treatments for aggression.

But, Yudofsky said, there will be opposition to this drastically different approach. Just as some people in the past wanted mentally ill people to be branded as evil, today there are many interest groups who do not want to reconceptualize violence as having a biological foundation, he said. "The brain is left out of the whole paradigm in the criminal justice system," he said. "We got nowhere punishing mentally ill people, and we're getting nowhere with our population of criminals. We're just building more prisons."

7 How the Brain's Chemistry Unleashes Violence

The first person to push a button when a light flashed got to punish his partner with a charge of electricity. The winner could choose a charge ranging from 1, a mere tingle, to 8, a painful jolt.

The subjects—normal, healthy students at McGill University in Montreal—usually gave tit for tat, making their partners endure no more pain than they received.

Then, to demonstrate the impact that certain brain chemicals have on violent behavior, the scientists in charge of the experiment turned up the volunteers' aggression levels.

The students drank a special concoction of amino acids that lowered their levels of the brain chemical serotonin. Soon, the volunteers began pushing numbers above 4, inflicting more pain on their partners, despite receiving below-4 charges themselves.

Then the students were given a large dose of tryptophan, which the brain uses to make serotonin. As the serotonin levels rose, the students became less aggressive, decreasing the amount of punishment they gave their partners, regardless of the increased pain their partners gave them.

The experiment's finding—that aggression can be controlled by manipulating brain levels of serotonin—was a profound discovery. For the first time, the raging monster was being forced into the open where scientists could begin to understand and tame it.

"What this is telling us is that low serotonin is involved in aggression in humans," said Simon Young, a behavioral neurochemist at

McGill, who is conducting the ongoing experiment with psychologist Robert Pihl.

"It's consistent with the idea that some of the people who become aggressive with alcohol are those who have low serotonin, and it supports the idea of using tryptophan in the treatment of aggression," he said.

Some other emotions besides aggression also were affected by Young's experiment. The volunteers said their moods changed and they felt a little more depressed when their brain serotonin levels were depleted. Their moods brightened when serotonin flooded back.

In earlier studies Young showed that alcohol could also lower serotonin and increase aggression, and that large doses of tryptophan could quell aggression in violent criminals.

Although serotonin isn't the only neurotransmitter linked to aggression, it plays a key role in determining how a human, or an animal, reacts to different situations.

Serotonin enables our drives to live in harmony. From their home base in the midbrain, each serotonin-producing cell sends out as many as 500,000 connections to cells in every part of the brain—the only neurotransmitter that does so—keeping the peace with gentle, rhythmic pulses of serotonin.

"The fact is that even if your serotonin is low, it doesn't compel you to be aggressive," said Dr. Emil F. Coccaro, director of clinical neuroscience research at the Medical College of Pennsylvania in Philadelphia. "It just permits you to be more aggressive, it lowers the threshold," he said. "You just don't stop yourself."

When scientists tested the first serotonin-lowering drug in rats, they thought the drug was an aphrodisiac because it unleashed the rat's sex drives and the animals copulated madly. They soon discovered, however, that the drug also unleashed aggression. The rats, which normally did not bite, became vicious, biting their handlers and attacking anything that came near them.

So far, sixteen different serotonin receptors have been found on brain cells and more are expected to be found. The wide assortment of receptors—tiny doors on cell surfaces through which neurotransmitters enter—explains how serotonin can affect so many drives and emotions.

If serotonin acts as a brake on impulses, noradrenaline, the brain's alarm hormone, acts as an accelerator.

"If people are leaving you alone and your brakes are lousy—your serotonin is low—it's okay. You're not going to get into trouble because your car is moving very slow," Coccaro said.

"But if people are starting to agitate you and your accelerator starts to rev, and your brakes are bad, you can't stop yourself. You lash out."

Scientists call the ability to restrain aggression impulse control, and the braking power a person has depends on both environmental experiences and genetic inheritance.

Many people inherit a gene that makes them more susceptible to low serotonin. But early life experiences—living in a violent household or a normal one—appear to determine how that gene will be expressed; that is, whether serotonin levels will be set on low, normal, or high.

Low levels seem to be an adaptation to a threatening environment. Low serotonin allows an individual to be more impulsive, and among the primitive drives that can then go unchecked is aggression.

Aggression increases a person's chances of survival under hostile conditions, but low serotonin levels also increase the risk of being aggressive at the wrong time and engaging in criminally violent behavior.

"If an individual lives in an environment that exposes him to a high risk of damage or death, then it would be a potential survival characteristic to be more impulsive, to get out of the way, or to respond to a threat quicker," said Dr. Robert Trestman, a psychiatrist at Mt. Sinai School of Medicine in New York.

While low serotonin levels increase impulsiveness, normal levels are associated with clear thinking and social success. With just the right amount of serotonin, the brain musters all of its resources to make use of the opportunities in its environment, balancing the risks against the benefits. In such individuals, enough aggression is allowed to surface so that they can be assertive and get things done. These are the leaders and achievers.

Dr. Michael Raleigh of the University of California at Los Angeles showed in male vervet monkeys that he could turn some of them into leaders and then convert them back into followers by raising or lowering their serotonin levels.

When the dominant monkey in a colony was removed and a subordinate male was given a drug like Prozac to increase serotonin, the monkey took charge. But the newly vitalized male did it in a clever way, by making friendships and alliances with the female monkeys. The new leader formed bonds with other colony members and then used these bonds to get the group to act together to vanquish its opponents.

Intrigued by his findings, Raleigh decided to study serotonin levels and the pecking order in college fraternities. Sure enough, those members with the highest ranking and the most friends had serotonin levels about 20 to 40 percent higher than members whose ranking was much lower.

While the monkey leaders who ruled through democracy had high-normal serotonin, another study found that some monkey leaders with below-normal serotonin ruled like dictators.

The low-serotonin leaders ran their group by beating up whoever got out of line, while the higher serotonin leaders made friends with the other monkeys and won them over to their side through affection and cooperation, said developmental neuroscientist J. Dee Higley of the National Institute on Alcohol Abuse and Alcoholism, which funds Raleigh's studies.

Normal serotonin levels are important for social development, Higley said. "One of the things you have to do to get along in modern society is take turns. Taking turns is the ultimate in impulse control. The reason that aggressive individuals have difficulty getting along is that they don't withhold their impulses."

Levels of serotonin that are way above normal—a response to an overwhelming environmental threat—have the opposite effect of low levels. The brakes lock. The brain is literally stopped cold, afraid to do anything, like an animal that freezes when confronted by a superior predator.

In humans, high serotonin is linked to the kind of fearfulness and rigidity of action that is seen in obsessive-compulsive behavior.

"These are people who are encumbered by anticipatory anxiety," said Dr. Paul Andreason of the National Institute on Alcohol Abuse and Alcoholism. "They think something is going to happen that's bad." He said such people often check and recheck doors to make sure they are locked, or repeatedly wash their hands to get rid of imagined germs.

Using PET scans to measure the brain's activity, Andreason found that obsessive-compulsive patients have excessive activity in the front part of their brains. Since this is the part that controls rational thinking, overloading it appears to knock out its ability to impose reason on basic drives.

In low-serotonin patients with lifetime histories of aggression, he found the reverse—low brain activity. The brakes weren't working.

The first clue that human aggression and violence were influenced by brain chemistry occurred in 1976 when Marie Asberg, at the Karolinska Hospital in Stockholm, documented a link between low serotonin and violent suicides.

These involved people who killed themselves with guns, knives, ropes, or by jumping from high places. Asberg found that those with low serotonin levels have a tenfold greater risk of violent death than other equally depressed patients with higher serotonin.

Her findings were met with disbelief and hostility. Fcw people thought that there could exist any correlation between a brain chemical and something as personal as suicide, Asberg recalled. "For most people, ending your days is more of a philosophical question and a personal question; not a biological question," she said.

The next major development came in the late 1970s and early 1980s when Gerald L. Brown and Frederick Goodwin of the National Institutes of Health showed that people who had long histories of criminal violence also tended to have low serotonin.

A look at their childhoods showed troubled behavior, repeatedly getting into fights, lying, using offensive language, setting fires, and killing their pets in a fit of rage.

Both suicide and a history of criminal violence are tied together by low serotonin. Why? They seem to be opposites.

An answer to the riddle began to emerge with noradrenaline, the hormone that alerts the brain to act in emergencies. When high noradrenaline was superimposed on low serotonin, impulsive aggression was aimed at others. When low noradrenaline was combined with low serotonin, aggression was aimed inward.

Neuroscientists had found a possible biological explanation of Sigmund Freud's observation nearly a century ago that anger turned toward others is aggression and anger turned against the self is depression.

Then, in the late 1980s, Dr. Markku Linnoila, scientific director of the National Institute on Alcohol Abuse and Alcoholism, found that the kind of violence associated with low serotonin is impulsive, hot-blooded, involving a loss of control.

This behavior characterizes irritable people who fly off the handle at the smallest challenge or perceived provocation—assaulting people, setting fires, or committing violent suicides.

Linnoila's study of 1,043 New York arsonists, those who did not set fires for profit, showed they had very low serotonin. Another study of 58 prisoners convicted of manslaughter predicted with an 84 percent degree of accuracy those who went out to kill again after their release—they had low serotonin.

A disturbing feature of these studies was the greater risk that children of violent fathers had of becoming violent themselves, suggesting a genetic link to aggression.

To test it, Linnoila and his colleagues divided 274 baby rhesus monkeys into three groups to create different environmental stresses. The monkeys were sired by males whose aggressive behavior ranged from low to high.

In one group the babies were raised with peers, the equivalent of being in a gang and separated from parental influences and socializing behavior. The second group was raised by unrelated nursing females to mimic a neutral upbringing, and the third group was raised by their own mothers, who looked after their every need.

Despite the three drastically different upbringings, the thing that most influenced the serotonin levels of the young monkeys was the genes they inherited from their fathers. Monkeys whose fathers were aggressive also tended to be aggressive even though they were raised in the most benign and loving group. Similarly, monkeys whose fathers were gentle, tended to be less aggressive despite growing up in a ganglike environment. Linnoila found that genes accounted for 60 to 70 percent of the aggression linked to low serotonin levels.

Surprisingly, when he went to search for "aggression genes" with the newly available tools of molecular biology, he discovered that a single gene defect was implicated in low serotonin.

The gene directs the production of a protein called tryptophan hydroxylase, which converts tryptophan from such foods in the diet as wheat and other grains into serotonin.

An error in the gene appears to make its possessor vulnerable to producing too little serotonin, probably under stressful conditions, such as living in a violent family or drinking too much alcohol, Linnoila said. "You need two hits. You need the mutated serotonin-making gene and you need a stressor, like alcohol."

Alcohol has two major effects on the brain. It initially raises serotonin levels so that a person feels more mellow for a brief time. Some scientists believe that this is why many aggression-prone individuals and depressives self-medicate with alcohol.

But then the nasty side of alcohol quickly shows up. Continued drinking precipitates a drop in serotonin, and for those with the mutated gene their levels can fall to the point where impulsive aggression breaks out.

Without the second hit, people with the flawed serotonin gene can live normally, even becoming very successful, and never get into trouble because of explosive outbursts. But their impulse to aggression lies just below the surface. Although they don't hit others, normal people with low serotonin are more prone to verbal acting-out and hostility when stressed.

The mutation appears to be common, showing up in 40 percent of the Swedish population tested at random. Further studies showed that almost all of the violent groups tested—violent alcoholics and criminals and those with antisocial behavior—possessed the mutated gene.

Alcohol is linked to violent criminal activity and the reason may be because it causes the mutated gene to make less serotonin, Linnoila said, adding that 48 percent of homicides in the United States are committed under the influence of alcohol.

Most striking were people who attempted suicide more than once. Of 100 such people, all had the defective serotonin gene.

"This was a surprise," said Linnoila. "I don't think we or anybody else expected a single gene to explain multiple suicide attempts."

ALCOHOL AND AGGRESSION

Alcoholism and other mental disorders may be genetically linked to a tendency towards violent activity. The link may be a mutated gene that causes reduced production of serotonin in the brain. Serotonin acts as the brain's brakes, making sure that basic instincts and emotions such as aggression don't race out of control. Initially, alcohol raises serotonin levels, but continued drinking causes a drop in serotonin and an increased risk of aggression. The following charts illustrate the probabilities of violence in male and female risk groups:

▶ PROBABILITIES OF VIOLENT BEHAVIOR

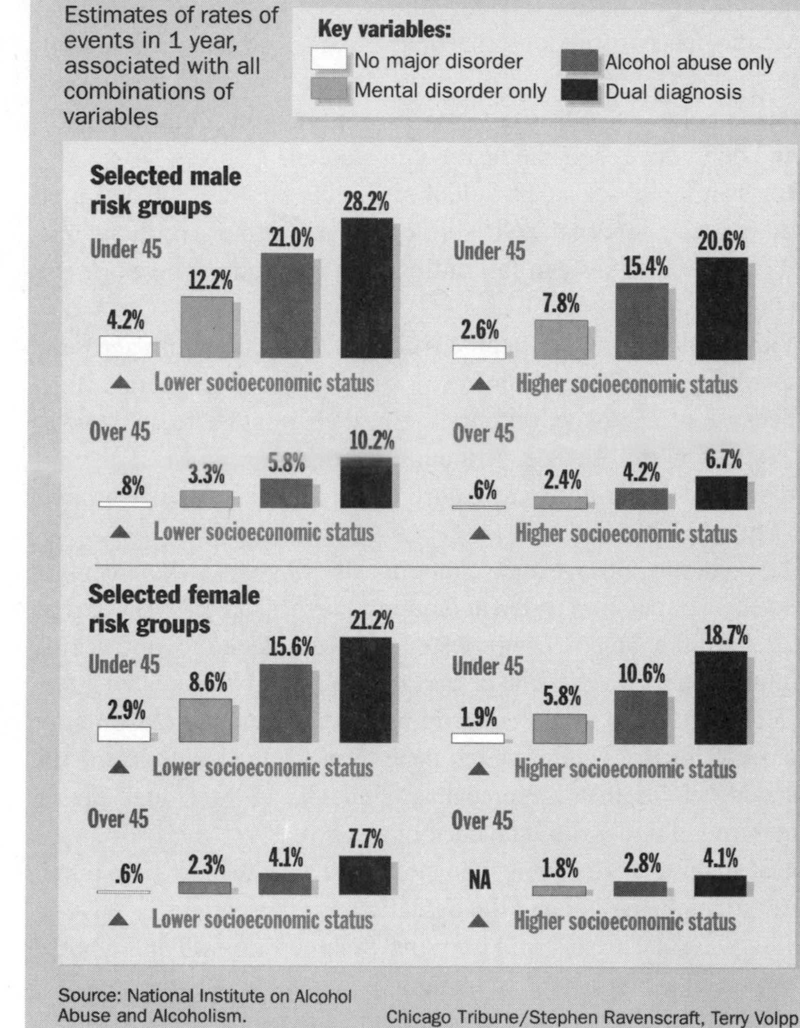

Source: National Institute on Alcohol Abuse and Alcoholism.

Chicago Tribune/Stephen Ravenscraft, Terry Volpp

For many people with low serotonin whose lives are a storm of aggression there is a glimmer of hope, Linnoila added. Aggressive sociopaths tend to calm down in their forties, primarily because their serotonin levels go up with age.

Another piece of the violence puzzle fell into place in October 1994 when Hans G. Brunner of the Netherlands reported another aggression gene.

This gene makes an enzyme responsible for breaking down noradrenaline and some other neurotransmitters so that they do not accumulate to dangerous levels. Brunner found a mutated form of the gene in certain families that is less efficient at destroying noradrenaline, the "accelerator" chemical. As the defective gene allows noradrenaline levels to climb, the gas pedal gets pushed down, the brain races, and the risk of unprovoked, uncontrolled rage increases.

The noradrenaline enzyme is called monoamine oxidase A, and it was found in a family in which the men were prone to violent outbursts. Some were arsonists, some rapists, and some tried to kill people with little provocation.

The defective noradrenaline gene is carried on the X sex chromosome. Only males show symptoms of the aggressive disorder because they inherit only one X chromosome along with a Y. Females always have two X's, one of which contains a good gene that overrides the bad one.

Other aggression genes are certain to be found. René Hen of INSERM, a research institute in Strasbourg, France, announced one such gene in animals.

Using genetic engineering techniques to deactivate a gene in mice that makes one of the sixteen known brain ccll receptors for serotonin, Hen reported that he was able to make the mice twice as aggressive as normal animals. The receptor, called 5HT-1B, occurs in different parts of the brain, including the primitive limbic system, which plays a major role in emotions.

"Our discovery that the 5HT-1B receptor might be a putative 'crime gene,' suggests that drugs acting at this receptor might be used to decrease criminal or violent behavior," Hen said.

The biology of violence is becoming clearer, said Dr. Dennis Charney of Yale University Medical Center. Like other behaviors, he added, aggression is moderated by genes and environment. "Your parents pass on a whole variety of things to you, and that includes your vulnerability to depression, anxiety, panic, psychosis, impulsivity, and aggression, all of which are influenced by serotonin," Charney said.

Charney showed he could precipitate depression in people prone to the mood disorder by lowering their brain serotonin levels, even if they were on antidepressive medications.

"So if you're brought up in a terrific environment, maybe aggression doesn't get expressed," he said. "If you're brought up in a very stressful environment, in which violence is part of everyday life, it may come out."

Social stressors are the key, said Linnoila. The economic benefits in American society are unevenly distributed among ethnic groups, increasing the likelihood of untoward behavioral outcomes in some of them.

"It is very important to realize that the results of these studies apply to the human condition in general and not to a particular ethnic group. Those groups of people who are most adversely affected by violence should benefit most in the end when all the results are in," said Linnoila.

8 Why Some Kids Turn Violent

In probing the biology of violence, scientists have found it useful to take an age-old question—Why do some kids turn out bad?—and pose it this way: What happens inside a developing brain to turn a child into a killer?

Their discoveries are shedding new light on the epidemic of violence that is being inflicted on children and that many of these children are inflicting on others.

Consider the infant brain: Its main job is to figure out the kind of world it will have to live in and what it will have to do to survive.

For millions of American children, the world they encounter is relentlessly menacing and hostile. So, with astounding speed and efficiency, their brains adapt and prepare for battle. Cells form trillions of new connections that create the chemical pathways of aggression; some chemicals are produced in overabundance, some are repressed.

What researchers now can tell us with increasing certainty is just how the brain adapts physically to this threatening environment—how abuse, poverty, neglect, or sensory deprivation can reset the brain's chemistry in ways that make some genetically vulnerable children more prone to violence.

The research also has produced an unexpected and ominous revelation: Environmentally induced brain changes can become permanent in an individual, encoding into genes a propensity for aggression and violence that can last a lifetime. These genes are not passed on to children because the genetic changes occur in brain cells, not in sperm cells.

But a tendency toward aggression can be inherited. Scientists have found that aggression genes, those that raise a person's propensity for violence, may be passed on to new generations. Some researchers believe that the increase in female criminal violence since the 1950s may be an early sign of how the genes of violence already are building up in the population.

"Aggressive or violent behavior has to be explained in part by biology," said Dr. Burr Eichelman, chief of psychiatry at Temple University in Philadelphia. "Even though there are all kinds of social and learned issues that get played out [in violent behavior], they are all superimposed on a biological substrate."

At the University of Illinois Medical School on Chicago's West Side, researchers are examining the blood of children for low serotonin levels. They know from an earlier study that these children are likely to grow up to be troublemakers and they want to find out why.

The study is designed to find out at what point in childhood serotonin levels plummet and what things in a child's early experiences cause serotonin to fall, said Dr. Markus J. Kruesi, chief of child and adolescent psychiatry at the school's Institute for Juvenile Research.

He already has done a study of twenty-nine children and adolescents with disruptive behavior disorders, which found that a low serotonin level was the single most accurate predictor of which youngsters would go on to commit more violent crimes or suicide.

"What we are all beginning to conclude is that the bad environments that more and more children are being exposed to are, indeed, creating an epidemic of violence," Kruesi said. "Environmental events are really causing molecular changes in the brain that make people more impulsive. It is frightening to think that we may be doing some very dreadful things to our children."

Other researchers are documenting the effects of bad childhood experiences on the brain's production of noradrenaline. Children who were raised in the Branch Davidian cult provide a vivid example.

Released during the siege of cult headquarters in Waco, Texas, in 1993, the children were found to have such high noradrenaline levels that their hearts roared in their chests, even at rest. While seated, the children had heart rates of 100 to 170 beats a minute. The average for children that age is 84.

Their brains, too, were racing, pumping out noradrenaline and other stress hormones in response to their violent and abusive lives with cult leader David Koresh, researchers said.

Dr. Bruce Perry, of the Baylor College of Medicine in Houston, who examined eleven of the Branch Davidian children after their release from the Waco compound, had seen the phenomenon before in inner-city Chicago children. The high levels of noradrenaline were the chemical signature of post-traumatic stress disorder, or PTSD, which first came to prominence among battle-scarred Vietnam veterans.

In children, the disorder resets the brain's chemistry to an alarm response. They are hot-blooded, more quick to react, more impulsive, more aggressive, and more likely to commit violent criminal acts.

"It is adaptive to be impulsive in that [abusive] setting," Perry said, referring to the cult. "If you wait, very frequently you will be victimized. So it's highly adaptive to be hypervigilant, to be overly reactive and impulsive, to actually act before you're acted upon."

Scientists now believe that along with the nation's increase in violence, low serotonin may be responsible for a steady increase in depression, especially among children. Since World War II each succeeding generation has had higher rates of depression, particularly among adolescents and young adults.

"Children born in later years are more likely to become depressed and to become depressed at earlier ages," said Dr. Peter Jensen, chief of the National Institute of Mental Health's child and adolescent disorders research branch.

He said the collapse of the two-parent family, poverty, teenage pregnancies, and violence are exposing many of these children to enormous stress.

"We have earned the dubious distinction of doing less for our families and children than any other industrialized nation," said Cornell University developmental psychologist Urie Bronfenbrenner.

Reported cases of physical abuse, sexual abuse, neglect, and emotional maltreatment of children jumped from 416,033 cases in 1976 to 1.7 million in 1990, according to a recent study by the National Academy of Sciences' National Research Council.

The rising tide of abuse and neglect of children occurs during the critical period when children are developing what Harvard's Felton Earls calls "moral emotions."

These are emotions that are rooted in brain chemistry and are established in the first three years of life. The development of impulse control occurs at a time when sensitivity toward others is also being rooted in the child's personality.

The chemical patterns that are established tell a child how to react to his or her environment, whether the child sees the world as a hostile place that has to be fought, or a more peaceful one where social cooperation wins the day. "We have a very naive belief that families are providing children with all their needs in these early years, despite the fact that families are functioning less and less well all the time," Earls said.

In 1992, for instance, of the nearly 66 million U.S. children under the age of eighteen, nearly 20 million lived with one parent. Among African-American children, 64 percent lived with a single parent, as did 35 percent of Hispanic children and 23 percent of whites.

For many children, a single-parent family is a war zone. Analyzing national child-abuse surveys, Dr. Richard Gelles of the University of Rhode Island found that severe violence toward children increased 71 percent among single mothers compared to mothers in two-parent homes.

Several trends contribute to the pattern of neglect and violence, said Dr. Markku Linnoila, scientific director of the National Institute on Alcohol Abuse and Alcoholism.

Large groups of disadvantaged people are thrown together in public housing complexes, where the physical layout and social patterns can make behavioral control over children difficult, he continued. "Then we have perhaps the most violent TV anywhere in the world. TV becomes the baby-sitter and the conflict-resolution patterns seen by the kids are blowing away the other guy. Then we provide the easiest availability of handguns, even automatic guns, in the world."

An estimated four million American children are victimized each year by physical abuse, sexual abuse, domestic violence, community violence, and other traumatic events.

Taking all of this in is the locus coeruleus, the brain's alarm network. Sitting at the base of the brain, it sends out noradrenaline pathways to other brain centers that control heart rate, breathing, blood pressure, emotions, and motivation.

When the locus coeruleus finds itself in an uncontrollable, threatening environment, it sets its noradrenaline gauge on high. Over the pathways come surges of the stress hormone that keep the body in a constant state of readiness—heart racing, blood pressure high, easy to startle, quick to blow up. These are the PTSD children.

They're in double trouble, said Baylor's Perry. They're at risk because they don't have the opportunities to learn the traditional ways that enable people to get along in society, and they are at risk because the brain systems involved in impulsivity are poorly regulated. The increased tendency to act before thinking, combined with language handicaps and poor problem-solving skills, predisposes these children to use aggressive and violent strategies to deal with life's daily challenges.

Setting the noradrenaline gauge on high is surprisingly easy to do. University of Wisconsin psychologist Mary Schneider found that noradrenaline levels can be turned up in a fetus if its mother is exposed to significant stresses during pregnancy.

Monkeys born to mothers who listened to ten minutes of random noise each day during mid- and late pregnancy had higher noradrenaline levels than normal monkeys. The hyped-up monkeys were impulsive, overresponsive, and had fewer social skills as infants.

When the prenatally stressed monkeys got to be the equivalent of preteens, their noradrenaline was still high and their behavior still abnormally hostile and aggressive, said University of Wisconsin psychologist Susan Clarke, who studied the monkeys as they grew up.

Although human studies are sketchy, they suggest that mothers who report high levels of stress during pregnancy, such as divorce, poverty, or violence, tend to have babies who are hyperactive and developmentally delayed. "If you think about the fact that the inner-city population is chronically stressed, and there's a lot of that population that is chronically pregnant, then we can begin to see some of the biology that may be responsible for high rates of aggression in the children," Clarke said.

While high noradrenaline and low serotonin appear to be behind the huge rise of impulsive, hot-blooded crime in the United States, scientists are also starting to study the effects of low noradrenaline, which is linked to cold-blooded, premeditated crime. Researchers are finding that the difference between high and low noradrenaline may be two sides of the same coin. Low arousers tend to be sensation-seekers. They try sky diving, auto racing, or other exciting activities to get their kicks in a socially acceptable way.

But some of them turn out to be more like Hannibal the Cannibal in the movie *Silence of the Lambs,* remorseless serial killers. More typically they are the kinds of criminals who coolly look for vulnerable victims they can stalk to rape, rob, or kill.

One study, using a PET imaging device to scan the brains of twenty-two convicted California murderers, tended to confirm the idea that brain chemistry determines whether a criminal will be a cold-blooded or a hot-blooded killer, said University of Southern California psychologist Adrian Raine.

"We broke the murderers down into those whose murderous acts were impulsive, where there was a lot of rage going on, and those who planned their acts, who were predatory," he said.

PET scans were used to measure chemical activity in the prefrontal cortex, the part of the brain that controls the expression of emotions. The scans revealed that the brains of the cold-blooded murderers were underaroused, suggesting low emotional activity, while the brains of the hot-blooded murderers were overaroused, indicating impulsive behavior.

Where do the cold-blooded killers come from?

Baylor's Perry is studying children who start off in the alarm state with high noradrenaline and impulsive behavior, and then around puberty convert to low noradrenaline, low arousal, and predatory behavior.

One explanation for the change may be that brain cells exposed to constant stress burn out, dropping to a lower level of activity to save themselves. Animal studies show that overexposure to stress can kill brain cells.

"It's really scary to watch the transition from high arousal to low arousal," Perry said. "They change from being victims to being victimizers and they develop this incredible icy quality of being emotionless." When they are arrested the only thing they are sorry about is being caught.

"If you ask them if there is anything they would do differently, the only things they would change are things that would prevent them from being caught, not the things that would prevent them from engaging in criminal behavior again," Perry said.

Psychologist David Magnusson of the Karolinska Institute in Stockholm has studied all the boys in a small Swedish town over a twenty-year period starting at the age of ten. He found that the few who go on to become career criminals have low noradrenaline levels. "The two most important characteristics of these persistent criminals are their lack of emotion and lack of remorse," he said.

Researchers have recently demonstrated that defects in two genes are linked to abnormal levels of serotonin and noradrenaline and that threatening environments can trigger these genes. Now they suspect that bad experiences can change genes and that those changes can quickly become permanent in the affected person.

"Sociologists have called me an idiot for saying this, because they think genes can't change in one generation, but regrettably they can," said Linnoila.

One of the common, everyday environmental influences that can directly affect genes is alcohol. It does so through demethylation, a process that cells use to upgrade or downgrade the activity of genes in response to environmental changes.

"There truly are novel mechanisms by which the environment can change genetic expression within a generation," Linnoila said. "What role it plays in our current epidemic of violence is not known right now, but the possibility that it does play a role cannot be ignored."

In his experiment to breed increasingly assaultive male mice, developmental psychobiologist Robert Cairns of the University of North Carolina in Chapel Hill noticed that their sisters were also becoming increasingly aggressive. Until that point, said Cairns, "I was a total disbeliever . . . [that] genes could change behavior."

THREATENING ENVIRONMENT AND AGGRESSION

The brain adapts physically to a threatening environment. Violence, poverty, neglect, harsh discipline, poor schooling, or sensory deprivation can influence the brain's production of two key chemicals implicated in violent agression and make some genetically vulnerable children more prone to criminal violence. The following selected charts highlight some of the environmental events that may contribute to the high rates of aggression in children:

CHILDREN LIVING IN POVERTY

Selected countries; 1991, in percent living below the poverty line*.

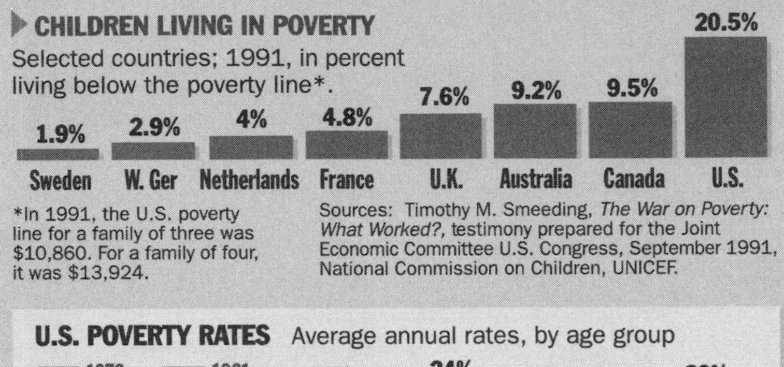

*In 1991, the U.S. poverty line for a family of three was $10,860. For a family of four, it was $13,924.

Sources: Timothy M. Smeeding, *The War on Poverty: What Worked?*, testimony prepared for the Joint Economic Committee U.S. Congress, September 1991, National Commission on Children, UNICEF.

U.S. POVERTY RATES

Average annual rates, by age group

	1979	1991
Under 6	18%	24%
Aged 6-17	16%	20%

Source: National Center for Children in Poverty.

Murder rates

Deaths by homicide per 100,000, 1987-90; average annual rates per year, age 15-24

	Deaths by homicide	Average annual rate
U.S.	5,718	15.3
Canada	121	3.1
Italy	179	1.9
Norway	9	1.4
Spain	93	1.4
Switzerland	12	1.3
Sweden	13	1.1
Denmark	8	1.0
Netherlands	21	.9
U.K.	80	.9
France	59	.7
Japan	73	.4

Suicide rates

Deaths by suicide per 100,000, 1987-90; average annual rates per year, age 15-24

	1970	1987-90
Austrailia	8.6	16.4
Norway	6.2	16.3
Canada	10.2	15.8
Switzerland	13.7	15.7
U.S.	8.0	13.2
Sweden	13.3	12.2
France	7.0	10.3
W. Ger.	13.4	9.6
Denmark	9.0	9.2
U.K.	4.3	7.2
Japan	13.0	7.0
Netherlands	4.9	6.7

Sources: World Health Organization, *World Health statistics Annual 1991.*

Chicago Tribune/Stephen Ravenscraft, Terry Volpp

It was amazing to see how genes for increased aggression could be transmitted to male and female offspring. The most aggressive offspring were successively bred to enhance the genetically aggressive line.

To find out if the same thing happens in humans, Cairns looked at the arrest records of teenage girls from 1900 to 1960 in Los Angeles County and then compared them with female arrest records in North Carolina during the '80s. Up until the 1950s, girls were arrested chiefly for sex offenses. In the 1980s their reasons for getting into trouble with the law increasingly involved violent crimes.

Like the generational increase in depression after World War II, every generation since the '50s also has seen an increase in young female aggression. The pattern that is emerging is of girls who are increasingly victims of child abuse, who grow up angry and have children with men who are likely to also be aggressive. As a result, succeeding generations of children are being born to aggressive parents and into aggression-inducing environments.

"It really suggests that if there are red signals that our society has to be wary of [they] should be those temporal increases in female violence. It has been ignored, but it may be the most important of all," said Cairns.

By heeding those red signals, scientists hope the new research on aggression and violence will open the doors to prevention and treatment.

The Epidemic of Violence

- Between 1950 and 1990 the homicide rate for white males aged 15–24 increased 4 times and that for black males 2.2 times.
- In that same period the suicide rate for white males increased 3.5 times and that for black males 3 times.
- Every day an estimated 270,000 students bring guns to school.
- Nearly one out of three children is born to an unmarried mother.
- One of every five children lives in poverty.
- One of every 50 children has a parent in prison.
- Alcohol is involved in 55 percent of all fights and assaults in the home.

Sources: Government and private child agency groups.

"If we don't invest in the early rearing environment of our children, we're going to be paying the bills for the rest of their lifetimes," said University of Wisconsin psychologist Christopher Coe. "The bills will be for mental disorders and physical diseases, and putting many of these kids in jail."

9 New Drugs That Fight Violence

Raw, primitive violence—violence without purpose—rules their brains. Almost every encounter with another person results in a fight. They are kept locked in the back wards of mental hospitals, feared and forgotten. Among them are intractable schizophrenics, murderers, rapists, arsonists, and plane hijackers.

Learning is impossible because of their explosive brain chemistry. Each new experience breaks into biological shrapnel before it has a chance to be laid down in memory.

Neuroscientists desperately have been seeking drugs to quell these violent explosions. As often happens in science, they have discovered much more than they set out to find—a whole new understanding of mental disorders.

As a result, drugs are being developed that not only quiet the raging brains of the most violent patients but also treat a wide range of other mental illnesses. Many of the drugs already are available by prescription, while some are administered only under strict supervision.

Unlike the standard antipsychotic drugs and tranquilizers, which often render patients dulled and sedated, the new medications leave them clearheaded.

One such drug is clozapine (Clozaril), which dampens explosive aggression and clears psychotic thoughts. At places like the Mendota Mental Health Institute in Madison, Wisconsin, clozapine has swung open the doors of the back wards, allowing patients once doomed to a lifetime under tight security to move into the community, going to school and work.

Doctors who have seen the drug's effect are enthusiastic. "It's like these people were living under a spell and clozapine is breaking the spell," said Dr. Gary J. Maier, of the University of Wisconsin, and director of psychiatric services at Mendota, which houses the state's most violent patients. "When that happens the long-standing immature personality that had been struggling to be healthy—but couldn't because it kept going crazy—is freed. They start to grow up."

Harvard's Dr. John Ratey, who treats Massachusetts' most violent criminals at Medfield State Hospital, also is sending some of his patients home after putting them on clozapine. He called it "the most exciting new drug I've ever seen" and likened its effect to "a guided missile that goes right to the site of aggression in the brain without making patients stupid, apathetic, sleepy, or nonsexual."

Clozapine and other new drugs such as sertraline, paroxetine, and buspirone were made possible by the discovery of the biochemical roots of aggression and depression, an imbalance of key neurotransmitters, such as serotonin and noradrenaline, and the receptors on cell surfaces that allow these chemical messengers in.

Scientists suddenly began to realize that they had stumbled upon a Rosetta Stone for decoding the mysteries of mental disease.

Low serotonin is common to many mental problems in which one or more of our drives bursts out of its chemical corral.

Medical researchers found that most of these disorders may be treatable with drugs that change serotonin levels. First developed to halt the uncontrollable aggression of schizophrenia and depression, these drugs are now being used or tested for a wide variety of problems, including alcoholism, eating disorders, premenstrual syndrome, migraines, anger attacks, manic-depressive disorder, obsessive-compulsive disorders, anxiety, sleep disorders, memory impairment, drug abuse, sexual perversions, irritability, Parkinson's disease, Alzheimer's, depersonalization disorder, borderline personality, autism, and brain injuries.

Low serotonin does not explain all of schizophrenia or aggression; other neurotransmitters like noradrenaline and dopamine play a role. But simply by resetting serotonin levels in the brain, as do clozapine, buspirone, and Prozac, a highly popular antidepressant, control over primitive drives and emotions can be restored.

One of Maier's patients, who averaged a fight a day and was kept

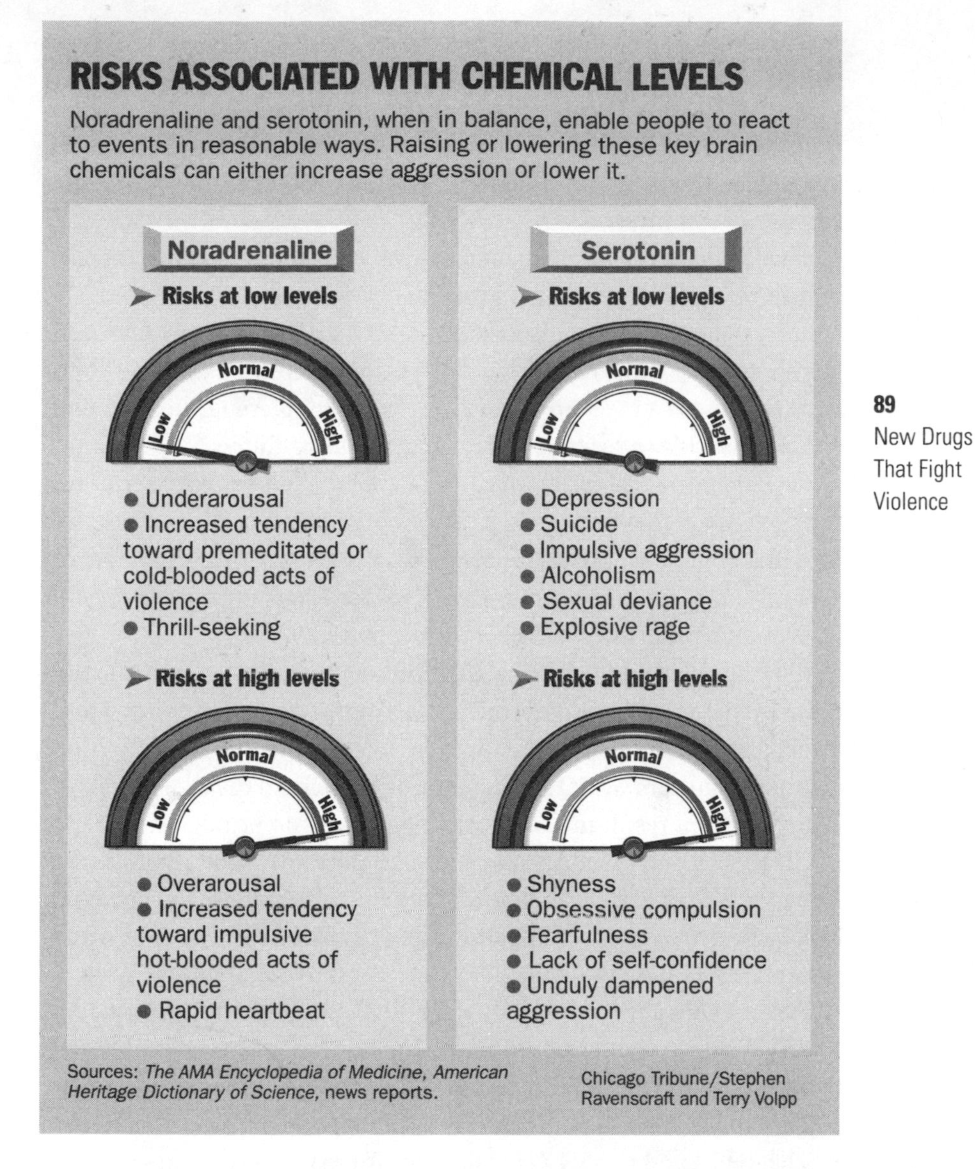

in maximum security for two years, felt as though a burden had been lifted after he started taking clozapine a year ago. "When his head cleared up and he stopped hearing voices that told him to kill himself, he became like a happy child. All he had to do was learn how to be an adult," Maier said. "We really just trained him. We taught him about responsibility and how to handle normal frustrations."

Because it seems to do so much with few side effects, Prozac has become a pharmaceutical star, with about ten million prescriptions written each year. But some mental health experts express concern that Prozac may be overprescribed and that it may be associated with more side effects than previously thought, such as insomnia, sexual problems, and jitteriness.

The fallout from aggression research is expected to have a major impact on almost all human behaviors and traits.

So far scientists have discovered sixteen different serotonin receptors in various parts of the brain, presumably modulating different basic drives and emotions including sex, appetite, pain, sleep, mood, and arousal. More receptors are expected to be found.

Yet, so new is the ability of drugs to open or close those serotonin receptor doors that scientists really don't know how they work, except that they do work. Some drugs work exclusively on specific receptors and others affect several. Still other drugs increase serotonin levels while others decrease serotonin output.

The idea behind drugs like clozapine is that they seem to restore a normal balance of brain chemistry. A single neurotransmitter whose levels are only 5 to 10 percent off normal can affect the way other neurotransmitters work, setting in motion a chain reaction of chemical errors that result in a wide variety of mental problems.

Clozapine, for example, may lock the doors of some serotonin receptors while opening the doors of others. Clozapine also appears to lock up some receptors of dopamine and noradrenaline, both of which are implicated in heightening the symptoms of schizophrenia.

Researchers are finding out what lies behind the receptor doors. The 5HT1 serotonin receptor, for instance, affects aggression, mood, and appetite and the 5HT3 receptor plays a role in learning and memory, and nausea and vomiting.

The 5HT2 and the 5HT1c doors are for psychosis, depression, alcohol abuse, and mood, while the 5HT6 receptor is for emotion and cognition and the 5HT1d is for migraine headaches.

Knowing the right serotonin receptor is helping scientists develop new drugs for conditions in addition to aggression. Two drugs recently approved by the Food and Drug Administration are ondansetron for nausea and vomiting and sumatriptan for migraines.

Prozac is also used to treat obsessive-compulsive disorder and the eating disorder bulimia.

Premenstrual syndrome, which affects 3 to 8 percent of women in their childbearing years, is now considered to be a form of depression linked to plummeting serotonin levels during certain times in the menstrual cycle. A study of more than 400 women suffering from PMS showed a major reduction in such symptoms as mental uneasiness, tension, irritability, and sadness when they were given small doses of Prozac.

Chemical messengers and their receptor targets not only regulate basic functions, they are programmed to work together. When a person becomes enraged, for example, his mood changes to anger, his blood pressure goes up, and his heart beats faster.

"All those kinds of reactions are supposed to be organized by higher brain centers that are involved in perceiving whether something is really dangerous or not and generating the best strategy to deal with it," said Dr. Steven Dubovsky of the University of Colorado Medical Center in Denver. "That's why when one goes out, the others go out. When regulation from above is out, impulsive urges take over."

Dubovsky is developing a computerized blood test he hopes will reveal the state of a person's brain by measuring neurotransmitter balances. Already he can measure chemical markers for twenty neurotransmitters, including serotonin, dopamine, and noradrenaline, and he is trying to develop patterns that signal mental troubles.

"We want to be able to develop biological profiles of people who are at risk for one kind of behavior or another, and then start looking for specific treatments for those profiles," he said. "We'll be able to get some sort of printout that would say 'Well here's the state of your receptors and here's the kind of balance between your neurotransmitter levels and here's the treatment you ought to get.'"

In a sense that already is happening. Drugs like clozapine and Prozac are undoing the violence-prone chemistry that is established in the brain as a result of stressful environmental experiences.

"The biology is not the cause of violence, the biology is in response to the way that children learn to become violent," said Harvard's Felton Earls.

The learning that goes on is very important in terms of how patterns are established in the nervous system as to what is important, what should be ignored, and what should be responded to.

"The wrong experiences set up the wrong brain chemistry and that manifests itself as low serotonin and high noradrenaline," Earls said. "For a lot of people, that is still hard to believe. This is something that we have not been able to make people visualize very well."

For all of its ability to halt aggression and clear minds, clozapine is not perfect. It helps only about half of the previously untreatable schizophrenics, and 1 out of 100 patients develops a life-threatening bone marrow disorder. Stopping the medication reverses the disorder.

But clozapine is also lifesaving. Counting on the drug's ability to halt impulsive violent aggression and to clear minds, Dr. Herbert Meltzer of the Case Western University Medical School in Cleveland has given clozapine to twenty people who have repeatedly attempted suicide with guns, knives, nooses, or jumping from high places. The drug reduced their suicide attempts by 60 percent over a two-year period. "We've definitely seen a marked decrease in aggression that is directed outward toward others or inward toward one's self," Meltzer said.

Prozac, the widely prescribed antidepressant, also has been found to work as an antiaggression drug, calming irritability and impulsiveness. Studies with twenty wife-beaters showed that Prozac reduced their aggression levels by 25 percent.

"They would say: 'My wife said something to me that would have really pissed me off before . . . I got kind of angry but I didn't feel like doing anything,'" said Dr. Emil F. Coccaro of the Medical College of Pennsylvania in Philadelphia, who has dedicated his career to finding ways to treat the biological causes of aggression.

Prozac seems to be giving wife-beaters and other impulsive aggressives more reflective delay, he explained. A person with normal serotonin takes time to think about a challenge, even if it's only a split second, to make sure he doesn't do the wrong thing.

"With Prozac they take time to size up the situation. We like to think serotonin does that for you naturally. It's the braking mechanism that prevents you from reacting to everything as soon as it happens," said Coccaro.

Reflective delay can also be increased through learning, he added. It works by teaching people to recognize when they are getting angry and either to walk away from an irritating situation or develop new problem-solving skills.

"We're going to need to come up with effective medications that increase the thresholds for acting aggressively and combine them with psychological interventions, like behavior modification, to give people longer fuses," Coccaro said.

The new serotonin drugs are surprising scientists in the kinds of basic drives they modulate. A Swedish drug called Amperozide was tested in pigs to determine if it could prevent their normal aggressive behavior. It was so successful that European farmers now use it to prevent aggression in their pigs, thereby avoiding the stress and meat loss that otherwise would occur. Amperozide is now being tested in humans as an antiaggressive agent, but scientists are also interested in the drug's potential for curbing alcohol and drug abuse.

Tests in animals show that a brief course of treatment with Amperozide in genetically alcoholic rats permanently reduced their desire for alcohol, the first drug ever to do so. No side effects, such as loss of appetite, were noted.

The compound, which seems to work on both serotonin and dopamine receptors, also reduces the craving for cocaine. Since dopamine is part of the brain's reward system, scientists speculate that Amperozide somehow gets the brain to recognize that a cocaine high is not a practical reward.

In a study conducted at the University of Chicago, Dr. Bruce Perry, now at Baylor College of Medicine in Houston, showed that lowering noradrenaline in children with post-traumatic stress disorder dramatically reduced their aggressive behavior.

Perry found that a drug called clonidine, which was originally developed to treat high blood pressure, also reduces noradrenaline in the brain, leading to major improvements in behavioral impulsivity, anxiety, arousal, concentration, and mood.

The ability to modify brain chemistry and change behavior with

CLONIDINE STUDY: MODIFYING BRAIN CHEMISTRY

Research in brain chemistry and the biochemical link to aggressive behavior has resulted in the use of drugs that not only help quell aggression and violence, but may be used to treat a range of other mental illnesses. The premise of these new drugs is that they seem to restore a normal balance of brain chemistry, leading to improvements in behavior, impulsiveness, anxiety, arousal, concentration, and mood.

- A University of Chicago study showed that the drug clonidine lowered the level of noradrenaline in children with post-traumatic stress disorder, which greatly reduced their aggressive behavior.

CLONIDINE STUDY

Four-week trial on children with severe chronic post-traumatic stress disorder. Psychiatric Symptom Assessment Scores (PSAS) represent means of assessing behavior, including aggression, at two weeks prior to and after the fouth week of clonidine treatment:

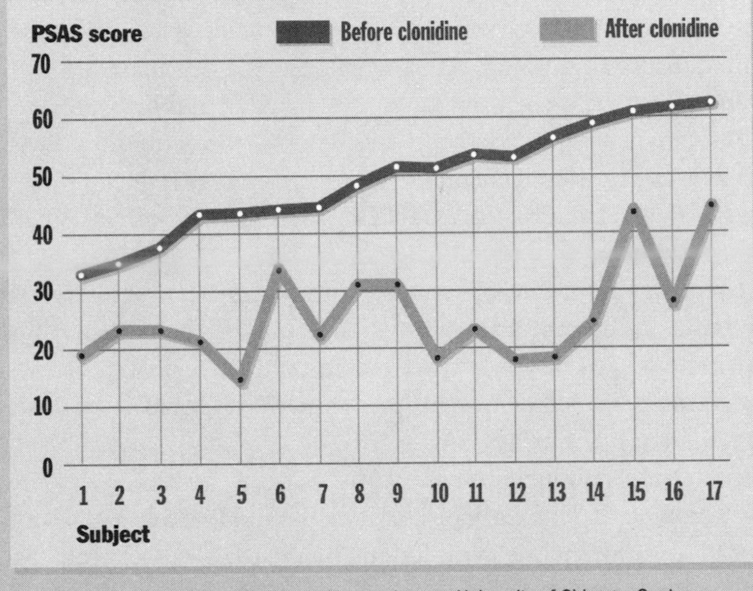

Sources: Laboratory of Developmental Neurosciences, University of Chicago, Center for the Study of Childhood Trauma,

Chicago Tribune/Stephen Ravenscraft, Terry Volpp

drugs also is having a fundamental impact on psychotherapy. Strategies like behavior modification or problem solving—techniques for unlearning harmful thinking habits in favor of good ones—are now believed to work by altering brain chemistry, much in the same way that drugs do.

Where Perry reduces noradrenaline with clonidine, child development psychologist Megan Gunnar of the University of Minnesota is doing it with a feeling of security.

Even children born with predispositions for high noradrenaline levels and fearful responses will remain calm in stressful situations if they are with a parent they trust. If the child does not feel secure with his caretaker, noradrenaline and hyperactivity go up.

"At every step, environment plays an important role in forming behavior. This is not an antibiological explanation. It is part of the biological explanation," Gunnar said. "Helping children to function more effectively in their social environment may be a very appropriate long-range strategy for preventing mental illness in general and violence in particular."

Harvard child psychologist Jerome Kagan sees this happening with shyness. About 15 percent of children, he estimates, are born with a shy temperament. But their shyness declines as they grow older and their environmental safety net helps them learn to overcome their inhibitions.

For every ten extremely shy children at the age of two, only five will be shy in kindergarten and first grade, three will be shy in adolescence, and by their twenties, only one or two will still be shy, Kagan said.

For parents of children who are uncontrollably aggressive and for whom psychotherapy and other drugs don't work, researchers at New York University and Bellevue University may have some good news. In a study headed by Dr. Magda Campbell, the researchers found that lithium, a compound used to treat manic-depressive disorders, and which acts on serotonin receptors, also quells the raging beast in these young people.

The drug was tested on fifty patients who were diagnosed as having conduct disorders, characterized by bullying, fighting, arguing, and temper tantrums. Forty percent of the children receiving lithium

GENETIC KEYS TO AGGRESSION

Scientists have found a biochemical link in which defective genes and the environment interact to produce abnormal levels of two mood-altering brain chemicals, serotonin and noradrenaline. Studies also suggest that threatening environments can trigger imbalances of these chemicals in people, laying the biochemical foundation for a lifetime of aggression or violent behavior.

➤ Where chemicals are found and what they do

Originating in the midbrain, these chemicals are known as neurotransmitters and enable cells in other parts of the brain to talk to each other by carrying messages between cells.

- **Serotonin:** The brain's master impulse modulator for all emotions and drives. It especially keeps aggression in line.
- **Noradrenaline:** The brain's alarm hormone. It recognizes danger and organizes the brain to respond to it by producing adrenaline and other chemicals.

Midbrain

Midbrain nerve cele

Nerve cell nucleus

DNA strand

➤ GENETIC DEFECTS

Two gene mutations have been found that support growing evidence of the genetic-environmental link to violence. One gene defect appears to lower serotonin and the other raises noradrenaline in susceptible people exposed to certain environmental stresses, such as violence and alcohol.

- Serotonin and noradrenaline may work separately or together in different abnormal combinations to produce a high-risk tendency toward a spectrum of violent activity.

➤ GENETIC INFORMATION STRUCTURES

Chromosomes: Structures within the cell nucleus that carry genetic information. Each chromosome contains a long strand of the hereditary substance deoxyribonucleic acid (DNA).

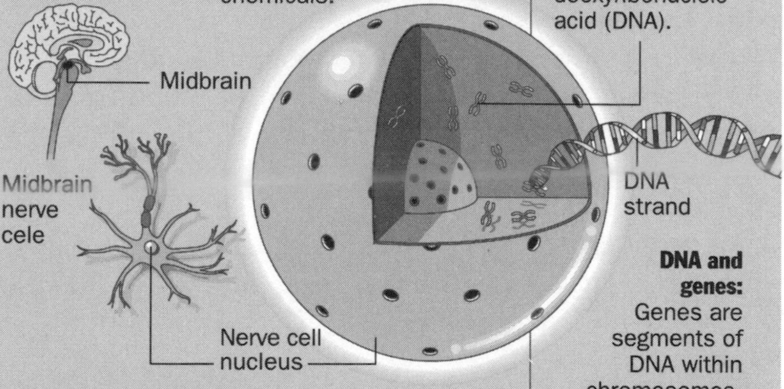

DNA and genes: Genes are segments of DNA within chromosomes. Each gene's function is encoded within the structure of its segment of DNA.

Sources: *The AMA Encyclopedia of Medicine, American Heritage Dictionary of Science,* news reports.

Chicago Tribune/ Stephen Ravenscraft and Terry Volpp

showed marked improvement. Only four percent of youngsters taking an inactive placebo showed similar improvement.

Food supplements may also be used one day to prevent aggression. Some researchers suggest that tryptophan, the compound that turns into serotonin in the brain, might be added to bread and other staples. The idea would be to make sure that the serotonin-depletion caused by drinking alcohol is counterbalanced with more tryptophan from the diet.

The Australian government successfully used a similar technique to reduce the number of cases of Korsakoff's psychosis. This brain disorder is caused by a depletion of the B vitamin thiamine in genetically susceptible populations, primarily those of British extraction, who are heavy drinkers. Australian officials mandated thiamine supplements in bread and other foods.

Self-medication with tryptophan may already be going on. Studies show that serotonin levels fall in women who experience severe premenstrual syndrome. Their craving for complex carbohydrates, including whole grains, breads, breakfast cereals, pasta, and chocolate, which are sources of tryptophan, may increase their serotonin.

"Through research we will be able to really delineate the major biological factors causing violent impulses and hopefully gain control of them," said Case Western's Meltzer. "It's quite likely that by the twenty-first century there will be a biology of temperament and character that can help us understand ourselves as a species."

10 How Alcohol Causes Craving

With astonishing precision, alcohol zips directly to the brain's pleasure center. For most people, that's not so bad. In fact there is a certain degree of benefit from a quick jolt of euphoria.

But for some people, alcohol, and a few other chemicals found in nature, can stealthily commandeer their brain's pleasure center, in effect robbing them of their willpower. They learn to crave something they were never supposed to crave.

Direct access from the outside to the brain's dispenser of pleasure, its reward system, was never intended by nature. It is too dangerous, a circumvention of the evolutionarily forged link between work and reward, a key to adaptation.

Within the last five years scientists have discovered that alcohol can stimulate the pleasure center in humans in at least five different ways. It works by altering the function of different chemical messengers in the brain that then produce alcohol's well-known effects—anxiety reduction, euphoria, sedation, disinhibition, aggression, forgetfulness, blackouts, tolerance, addiction, and withdrawal.

Not everyone has the same reaction to alcohol. And as familiar as some of alcohol's effects are, scientists have been unable to bridge the gap between its chemistry and its behavioral consequences until now.

All that is changing because of a revolution in science sparked by the powerful tools of molecular biology, which can probe the fundamental secrets of life, and amazing imaging devices that can see inside the body as never before.

Using one of those powerful tools, PET (positron emission tomography) scanners that can see the brain at work, University of Chicago scientists, for instance, have made the first images to show alcohol affecting the reward center in humans, and watched as alcohol went on to affect other areas of the brain.

The view is stunning, a biological pinball game, as first one section of the brain lights up, then another, then another. At the same time, other areas of the brain become dim as alcohol quiets activity in those cells. It's the first demonstration in humans of how alcohol, as it is metabolized in the brain, alters mood, producing changes that are totally predictable of the known pharmacological effects of alcohol, said University of Chicago radiologist Dr. Malcolm Cooper.

"We're at the interface between molecular biology and behavior," he said. "It's all rushing down on us in a superb way. We can now begin to relate behavior to genes and metabolism."

After consuming two to three drinks, a volunteer slides his head into the center of a PET machine, which measures how his brain is using sugar that has been marked with a radioactive tag. Sugar is the brain's chief source of energy.

The first thing alcohol does is to dampen the whole brain a bit, said Harriet De Wit, an experimental psychologist. This is related to alcohol's anxiety-reducing capability. But the next target that lights up, the reward system, generates a little stimulation, a little pleasure, a little euphoria.

John Metz, a University of Chicago research psychiatrist, likened the reward system, which sits in the midbrain and sends and receives messages from most other parts of the brain, to a traffic rotary. "Essentially it's a loop of anatomical connections," he explained. "If you interfere with this loop at any point you get a behavioral consequence on mood. And, depending on where you interfere with the loop, you get a different effect."

Analyzing the PET image of a person who had just consumed two to three drinks, Metz pointed out how alcohol perturbed the reward loop. The cerebellum, which controls basic motor function, and the visual cortex, toward the back of the brain, were now blue—the activity of cells in these areas was decreasing.

The limbic network in the midbrain, which is the heart of the reward system, was red, indicating that cells in this region were becoming more active from exposure to alcohol. The reward center was already telling the brain that it liked alcohol.

The red followed the path of alcohol-induced activity to the basal ganglia, involved in movement and cognition. From there it went to the left side of the brain where language is processed, and then up into the frontal cortex where reasoning and voluntary motor control reside.

"This is what happens when you drink," Metz said. "You get motor incoordination. You might have some visual distortions. You might slur your words. Your mood changes and your thinking processes are a little bit off. All of which are being shown in this image."

So far, the PET studies have shown three distinct reactions to alcohol, said De Wit. In one third of the subjects, the brain's reward system lit up when they drank; they got stimulated and they liked it. One third became sedated with alcohol; their reward systems remained unlit, and they didn't like to drink. The remaining third were somewhere in between and could go either way.

The reward system is designed to insure survival—not craving. The pleasure center is supposed to reward sex, eating, success, and other behaviors that enhance evolutionary adaptation. It makes people do things over and over again by giving them emotional boosters that range from intense highs to satisfaction for a job well done.

Normally the brain is stingy in doling out pleasure. People wouldn't be motivated to work if their brain kept them constantly supplied with wonderful sensations. If sexual behavior were not intensely but briefly rewarding, for example, the human race probably would not have survived.

"Rationally weighing the merits of sexual reproduction, absent some strongly motivating emotions, would clearly not have gotten the job done," said Dr. Steven E. Hyman, director of Harvard Medical School's Division on Addictions. "Without the brain's reward system that gives values, and therefore priority, to certain behaviors, it is indeed hard to imagine how complex-behaving animals could survive.

James Olds discovered the brain's pleasure center in the 1950s when he inserted a thin electrode into the primitive limbic system of rat brains. When the electrode was attached to a bar that could send a tiny charge of electricity to stimulate their pleasure centers, the rats pushed the bar nonstop, 5,000 times an hour, forgoing food, water, and sex, until they collapsed from exhaustion.

Compared to the electrode in rat brains, alcohol is a weaker stimulator of the pleasure center. However, some people are genetically vulnerable to more powerful stimulation from alcohol, leading to what is variously called compulsion, addiction, or craving.

Unlike other mental disorders, alcoholism can be created in animals, giving researchers a unique opportunity to study the behavior and brain chemistry involved in addiction.

One of the most successful animal models is a strain of alcohol-craving rats bred by Dr. Ting-Kai Li of the Indiana University School of Medicine, who is studying the genetics of alcoholism.

His drinker rats will press a bar to get a shot of booze. But unlike the Olds rats who pressed the bar thousands of times an hour, Li's rats press it thousands of times a day. One reason for their slower reaction time is that the animals eventually become intoxicated.

Craving is a discombobulating mental state that has puzzled, amazed, stupefied, and addicted many people since prehistoric times. But craving has always been a mystery. It has defied religious, legal, psychological, personal, and social attempts to control it.

Now it is coming under the close scrutiny of scientists, who are beginning to examine the biology of behavior, and they are making alcohol yield the secrets that give it the power to cause craving.

Surprisingly, they are finding that alcohol, the oldest and most prevalent cause of addiction, is by far the most prolific activator and deactivator of brain centers. Nothing else comes close. Not cocaine. Not heroin. Not nicotine.

Alcohol's secret is in its simplicity. It can be made from sugar water sitting in the sun lazily forming basic chemical molecules made up of oxygen, hydrogen, and carbon that have an unparalleled affinity for the brain.

Until recently, alcohol was thought to act like Drano, dissolving into brain tissue to create its toxic effects. It does that, but not in the way scientists thought. Alcohol dissolves in both water and fat, two of the main components of brain cells, selectively forcing these cells to talk to each other in ways that produce emotions and behaviors ranging from euphoria to depression and calmness to aggression.

"Alcohol is a lot more specific than we used to think it was," said George Breese, professor of psychiatry and pharmacology at the University of North Carolina in Chapel Hill. Breese headed a team of researchers who discovered in 1989 that alcohol picks its targets in the brain.

Like the brain's own chemical messengers, alcohol plugs into the brain's massive network of receptors, biological switches that normally turn cells on or off, enabling them to communicate with each other. But alcohol is devious. Because it can dissolve in water, it combines with water molecules that form part of the receptors. The shape of the receptor is thus changed, making it easier for a neurotransmitter to turn the receptor on or off, scientists believe.

In addition, alcohol affects channels in brain-cell membranes. These channels permit calcium and other electrically charged chemicals to enter cells to give them the energy they need to fire off a message to another cell.

Then the other devious side of alcohol—its ability to dissolve into fat—comes into play. By combining with fat molecules that form channels in the surfaces of cell membranes, alcohol temporarily changes their structure and their function.

Alcohol also appears to be able to seep into cells to prevent the messages imparted by neurotransmitters from being translated into instructions inside the cell.

To scientists, alcohol's ability to affect so many different parts of the brain is a paradox. Alcohol's effects are at the same time broad yet exquisitely specific, acting on one type of receptor in one part of the brain, for instance, but not on the same receptor in a different area of the brain.

"How is it that you have this substance that can in fact perturb the system so globally and not just shut it down totally," asked Herman H. Samson, professor of physiology and pharmacology at the Bowman Gray School of Medicine in Winston-Salem, North Carolina.

"When you drink, why don't you just drop into a puddle on the ground and that's the end of it?"

Answers to that question are now beginning to be found. Scientists are discovering how alcohol affects most of the brain, altering motivation, emotion, awareness, thinking, movement, breathing, consciousness, memory, and more.

"Alcohol research is on the threshold of making a giant leap forward in our understanding of the etiology of alcoholism," said Dr. Walter A. Hunt, chief of the Neurosciences and Behavioral Research Branch of the National Institute of Alcohol Abuse and Alcoholism.

A major part of that "giant leap" is the growing knowledge of the role that genetics plays in alcohol abuse. The genes that a person inherits from his or her parents may make that individual more prone or less prone to excessive drinking. For the most part, these genes work in the brain, doing such things as increasing a person's susceptibility to craving alcohol.

Scientists know there is not one "alcoholic gene." But there is growing evidence that alcoholism has a multigene basis. As with many other complicated disorders, a number of genes may interact with environmental influences—such as availability of alcohol, seductive advertising, friends who drink—to make a person vulnerable to the addicting effects of alcohol.

"There's been a controversy for more than forty years in this country as to whether alcoholism is a disease or whether it's a moral character disorder," Samson said. "I don't think there's any question in most of our minds that there are huge genetic components that are involved, which suggests that there's a subpopulation at high risk."

The goal is to develop tools that will be able to test for genetic susceptibility to alcoholism, and then measure the brain's reaction to alcohol through imaging.

"Eventually we may be able to screen people for the genes that may make them susceptible to alcohol abuse," said psychiatrist Dr. Henri Begleiter of the State University of New York's Health Sciences Center in Brooklyn. Begleiter is heading an NIAAA-funded study involving six centers in the United States and twelve other centers around the world in a search for the genes that increase the risk of alcohol abuse.

"People have a right to know if they are predisposed to alcoholism or not," he said. "Then we can develop early intervention programs in terms of diet, environment, education, or drugs to prevent dependence."

Scientists are also beginning to change their minds about whether abstinence is the only remedy for alcoholism. New treatments may one day be able to convert some alcoholics into safe drinkers. But for now, the safest approach is abstinence.

Because of the many things alcohol can do to the brain and the ease with which it can be made—from fermenting fruits or grains—its use dates back through all of recorded history and beyond, past the ancient Egyptians and Sumarians to the Stone Age.

Alcohol affects the body in ways that are both good and bad. On the good side is the growing evidence that moderate consumption—one to two drinks a day—is associated with better adult psychological health—because of alcohol's antianxiety properties—and a longer life. Moderate drinkers live longer than either teetotalers or heavy drinkers.

Increased life expectancy among moderate drinkers is due, in part, to their significantly reduced risk of heart attack, stroke, and possibly diabetes. Alcohol lowers LDL, the cholesterol linked to heart disease, and increases HDL, the good cholesterol that protects against clogged arteries.

Alcohol also prevents the formation of blood clots, which can cause heart attacks; makes cells more sensitive to insulin, thereby reducing the risk of diabetes; and increases estrogen production in postmenopausal women. It is when estrogen production dwindles after menopause that the risk of heart disease greatly escalates in women.

"On the basis of this and other information . . . clinicians probably should behave as if moderate alcohol intake is, on average, not just harmless but beneficial," Harvard Medical School's Dr. Robert H. Fletcher reported in the *British Medical Journal.*

For men, alcohol's potential benefits seem to accrue at one to two drinks a day. For women, it appears to be one drink a day.

A twelve-year study of more than 85,000 women recently found that light to moderate drinkers, those who consumed one to six drinks a week, lived significantly longer than either nondrinkers or heavy drinkers. Alcohol's main effect was to reduce the risk of heart disease in women over the age of fifty, concluded researchers from Harvard and the Dana-Farber Cancer Institute in Boston.

Should drinking be recommended as a national public health policy? No. At least not yet, and maybe never, said Dr. Enoch Gordis, director of the National Institute on Alcohol Abuse and Alcoholism. Despite its beneficial effects, alcohol still has the potential for causing harm in some people. The choice for each individual is a different matter, and probably should be made in consultation with a physician.

"If you want to start drinking to enjoy the pleasures of alcohol, and there are none of the risk factors, it's okay," he said. "If you're already drinking moderately, and you're not having any problems, it's okay.

"But, to drink solely for deriving cardiovascular benefits, that doesn't make sense. There are other things you can do to reduce your chance of heart disease, like diet and exercise."

Alcohol is used by 60.7 percent of Americans—about 72 percent of men drink compared with 51 percent of women. Most people drink safely. The average American drinker consumes the equivalent of two drinks a day.

The bad side of alcohol begins to emerge, however, as the number of drinks per day exceeds three. And it is well known that excessive drinking can cause mental, social, medical, and economic woes. Besides the brain, heavy drinking can harm the heart, liver, and other organs. Russian men, for instance, drink the equivalent of three and a half bottles of vodka a week and have a life expectancy of only fifty-eight years, fourteen years fewer than American men.

Given the fine line between benefit and harm, the question that has long perplexed scientists is why some people become dependent on alcohol and drink to excess.

Nearly 14 million Americans are dependent on alcohol. Their inability to control the amount they drink costs the nation more than $100 billion annually in medical care and lost work. One out of four hospital patients has an illness brought on by drinking. Alcohol accounts for one out of every twenty deaths in this country.

"One of the overwhelming statistics that should impress everybody is the fact that about 60 percent of all violent acts, whether murders, child abuse, family abuse, assaults, or felonies are associated with the consumption of alcohol," said psychopharmacologist Klaus Miczek of Tufts University. "If you want to make a dent in the violence problem, alcohol wouldn't be a bad place to start."

Since alcohol is the most commonly abused drug, scientists hope to use their new knowledge of its pathways in the brain to develop effective ways of preventing dependence. And they are making headway.

So far researchers have discovered ten different brain-cell switches that alcohol can turn on or off. Five of these switches turn on the brain's pleasure center, providing scientific evidence for the Book of Psalms' timeless observation that wine gladdens the heart of man.

But to modern-day scientists, the discovery of alcohol's ability to turn on the brain's reward system is the key to understanding how alcohol creates a craving so intense that it makes emotion rule over reason. When a person slips into dependence, alcohol craving becomes a drive as powerful as the need for food, water, sleep, and sex.

"It's an exciting time," said Samson. "We're on the threshold of understanding some of the mechanisms that regulate the intake of alcohol. We're not going to find a magic bullet that's going to cure alcoholism. But we will find better therapeutic treatments that will effectively treat this disorder."

11 Alcohol and the Brain's Reward Center

Surrounded by thick bone, the brain is the body's most protected organ. Untrusting of anything from the outside that might interfere with its complex and delicate operations, especially its reward system, the brain makes almost all of its own chemicals.

But nature created a flaw that has come back to haunt humans. It made chemical messengers in plants that can capture the brain's reward system and dispense its jealously guarded pleasures. This happens because once nature makes something that works well, it likes to reuse it in all types of flora and fauna, and it is the reason humans share many genes with other animals and plants.

And humans, being the most inquisitive creatures on Earth, would eventually stumble upon these mind-altering compounds.

Some of the chemicals that are found in fermentation (alcohol), coca and tobacco leaves (cocaine and nicotine), and poppies (heroin) are chemical cousins of the brain's own pleasure center messengers. As such, they can activate the center.

"It's by this very perverse serendipity that certain plant products are able to capture us," said Dr. Steven E. Hyman, director of Harvard Medical School's addiction division. "It's by virtue of the fact that these plant products mimic our neurotransmitters."

One of these pleasure switches recently yielded to the first new treatment for alcoholism in forty years and heralds a breakthrough for other new treatments.

Until now the only drug used to treat alcoholics was Antabuse, a compound that makes people sick when they drink. Newer drugs in development are designed to prevent alcohol from acting on brain-cell switches that activate the reward system.

The first one to do that involves a pleasure-center switch that normally responds to endorphins and other opioid-like neurotransmitters that produce a natural high.

Naltrexone, a drug already in use to prevent heroin from triggering the opioid pleasure circuit, also blocks alcohol's ability to switch on this circuit. Naltrexone doesn't decrease the desire for heroin, only its effect.

Surprisingly, however, naltrexone does block the craving for alcohol in some drinkers. As a result, their desire to drink is significantly reduced. The drug is now being used, along with psychotherapy, to successfully treat alcoholics.

Just as the brain makes its own opiates, which are used in regulating pain and stress, it also makes its own alcohol-like chemical.

"The exciting thing is that our brains have the capability of making a substance that acts just like alcohol when you give it to animals," said Kathleen A. Grant, associate professor of physiology and pharmacology at the Bowman Gray School of Medicine in Winston-Salem, North Carolina.

Scientists don't yet know what the alcohol-like compounds, called neurosteroids, do in the brain but they believe the chemicals may be involved in sleep and anxiety reduction.

The brain has two master switches, the "stop" switch, which tells brain cells not to fire, and the "go" switch, which tells cells to fire. The switches are technically known as receptors, ports on the surfaces of brain cells where neurotransmitters dock to deliver their chemical messages. The "go" switch is called the NMDA receptor and the "stop" switch is known as the GABA receptor. Together they oversee the function of the brain.

The "stop" and "go" switches are crucial because it is important for brain cells to know when to fire and when not to fire. Otherwise the brain would give the same weight to all experiences and learning would be impossible.

As scientists probe alcohol's biological activity they are uncovering the fascinating story of how our reward system works. The center is based in the limbic area, located in the lower midbrain, and it is capable of dishing out pleasurable as well as frightening emotions.

The limbic system is very primitive and it is sometimes called the lizard brain. The purpose of a reward system is to insure successful adaptation. All animals have one, even reptiles, which trace their ancestry to the foggiest evolutionary beginnings.

In humans, the limbic system connects to key parts of the brain, such as the hippocampus, where memories begin, and the cerebral cortex, from which higher thought emanates.

"The reward system brings limbic information—emotional information—into the area of the brain that makes you do things," said Dr. George Koob, a neuropsychopharmacologist at the Scripps Research Institute in LaJolla, Calif.

Performing like an emotional thermostat, the limbic system helps the brain sort though the zillions of sensory inputs and thoughts we experience each day by tagging them as emotionally hot or cold, or something in between.

Hot things arc to be remembered and they are spritzed with pleasurable sensations ranging from the thrill of sex to the glow from a good deed. Cold sensory inputs and thoughts receive no emotional tag and they drift away to be forgotten.

"The reward system affects behavior," Hyman said. "It says 'that was good, let's do it again and let's remember exactly how we did it.' In that way you learn important new things that are good for you."

The limbic system is also capable of branding experiences with negative emotions, such as fear. The part of the limbic system that imparts fear is called the amygdala. The amygdala also plays a role in reward, but one of its main jobs is to tell people what's frightening and should be avoided. It is here where the general dislike of snakes is rooted.

The remembrance of things that are really terrible, like child abuse or the terror of war, appears to be engraved into the brain through the amygdala. Sometimes the amygdala can suddenly resurrect these terrible memories in uncontrollable flashbacks. The flashbacks are triggered when a person experiences something that reminds him of the past terror, a condition known as post-traumatic stress disorder.

"Once Nature worked out ways in which we could learn about

things that were good for us and reward them, and things that were bad for us and make them scary, those things got held onto in evolution," Hyman said. "If you step on a snake once, you're not going to do it again. You learn very quickly to be scared of creepy crawlies. That would be a very important thing for survival." Early in evolution, nature figured out how to deal with these survival issues and they're still there in our brains.

By tracing the reward circuits, researchers are discovering what happens inside the brain when a person drinks to reduce anxiety or to forget.

Alcohol pushes the GABA "stop" switches on brain cells to shut the cells down and, at least early on, to induce a sense of calm. They are the same switches that barbiturates push to bring on sedation and anesthetics push to produce unconsciousness.

But as more alcohol is consumed, larger sections of the brain are turned off, ultimately producing anesthesia and loss of consciousness.

Alcohol has the opposite effect on the NMDA "go" switch. Most brain cells have a "go" switch that commands the cells to fire, thereby passing on information necessary for memory formation. Alcohol blocks the "go" switch, preventing cells from firing and new memories from being laid down.

So, when you can't remember the names of people you met at a cocktail party or you have blackouts, it's because alcohol kept your "go" switches from being turned on.

At that point, your brain is like a camera without film. The things you heard and saw the night before are blanks because those experiences did not get branded into your brain's circuitry. The "go" switch is also the one that is blocked to produce amnesia, and that when turned on too much can cause cells to become overexcited and produce seizures.

Both the "stop" and "go" switches are tied into the reward system. As a result, things done while drinking that involve these switches become tinged with pleasurable emotional memories. That's why an urge to drink can well up in an abstinent alcoholic when he passes the bar where he used to drink or sees an advertisement of his favorite beverage.

Together the "stop" and "go" switches set in motion the complex phenomenon of craving and the pursuit of alcohol even in the face of such adversities as hangovers, illness, divorce, and lost jobs.

"We're now using very sophisticated neurobiology tools to understand the different receptor types and how they react to different doses of alcohol," said Herman H. Samson, professor of physiology and pharmacology at the Bowman Gray School of Medicine. "Five years ago everyone thought that was a pipe dream. The field has advanced at such lightning speed that within another five years we're going to be able to answer some of those questions."

Besides the "stop" and "go" switches, three other receptor systems feed into the reward network to induce craving. In addition to opioid receptors, they include serotonin, a messenger that regulates such basic drives as eating and sex, and dopamine, which regulates emotion in one part of the brain and motor control in another.

Each of these receptor systems produces a different effect when alcohol is consumed, and people may be more susceptible to one effect than another.

Type 1 alcoholics, for example, react more to alcohol's tension-reducing effect on the GABA system. These alcoholics tend to be more anxious and usually become problem drinkers after the age of twenty-five.

Type 2 alcoholics, on the other hand, drink for alcohol's euphoric effect, which is regulated through the serotonin and dopamine receptor systems. "These people tend to be more impulsive and aggressive," said Dr. Emil F. Coccaro of the Medical College of Pennsylvania in Philadelphia, who has found that Prozac, which increases serotonin levels, significantly reduces aggression. "Alcohol dependence usually starts in their teens and they drink for the euphoric properties of the drug."

As the number of drinks increases, different parts of the brain react in dramatic ways, said Kathleen A. Grant, who has been studying how different doses of alcohol affect neurotransmitters, the brain's chemical messengers, in mice.

"At low doses of alcohol one neurotransmitter system might be mediating most of alcohol's effects on the brain," she said. "But as you continue to drink, those effects might be supplanted by another neurotransmitter system that was initially less sensitive to alcohol, then suddenly becomes highly active."

Based on animal research and human behavioral studies, here's what consumption of varying amounts of alcohol does to the human brain:

- One drink. Increases stimulation and motor activity.
- Two drinks. Decreases anxiety and susceptibility to stress and tension. Causes some euphoria and reduces inhibitions.

- Three to four drinks. Reduces stimulation. Produces early signs of memory impairment (noticeable as forgetfulness) and motor incoordination.
- Five to six drinks. May cause some amnesia or blackouts. Dulls reaction to pain and slows down breathing. Motor behavior becomes very slow and uncoordinated. Person becomes stuporous as alcohol's anesthetic qualities start to show.
- Seven or more drinks. May cause loss of consciousness.

Scientists know that all of these things occur in humans because they can prevent them with drugs that block specific receptors. As a result, receptors don't respond to alcohol and the reward system becomes temporarily immune to its effects.

"The genetic susceptibility to addiction runs through the reward system," said Harvard's Hyman. "It's like having your thermostat set a little bit differently so that drugs either light it up or you feel not quite so good until you use drugs. At the same time, they produce nearly indelible emotional memories that predispose the person to drug craving and hence to relapse."

One of the circuits, involving serotonin, is now thought to be a key to why many people start drinking.

At least sixteen different serotonin receptors have been discovered so far. They are involved primarily with impulse control, setting the levels of our drives for sex, food, aggression, and general well-being.

Alcohol can enhance many of those behavioral set points by increasing serotonin in some of the receptors. People with low serotonin often don't feel good about themselves and so they drink to feel better.

Scientists now believe that alcohol at first increases serotonin levels. Mood improves and people become more sociable. Indeed, Ting-Kai Li's alcohol-drinking rats at the Indiana University School of Medicine start off with low serotonin that is increased when they imbibe. The rats are aggressive before they drink but calm down and stop biting their neighbors as their alcohol intake increases.

But some researchers believe alcohol produces a rebound effect. After initially increasing serotonin, alcohol may cause it to drop as alcohol consumption continues, thereby increasing the risk of impulsive aggression and depression.

Dr. Markku Linnoila, scientific director of the National Institute of Alcohol Abuse and Alcoholism, found that the sons of alcoholics have lower serotonin levels, suggesting that low levels are an inherited trait.

Interestingly, Linnoila and his colleagues found that serotonin levels naturally increase as these men reach their forties and their aggressive behavior declines. They mellow out.

"Several of these people have related to us that they've stopped drinking," he said. "Alcohol doesn't please them anymore. They don't get the same kick, the same euphoria."

Researchers can also diminish the euphoric effect of alcohol in abusers by giving them drugs that increase serotonin. A number of studies are under way to test these drugs, called serotonin reuptake inhibitors, a major new class of antidepressants. The serotonin drugs may join naltrexone, which blocks alcohol's effect on the opioid receptors, in the treatment of alcoholics.

"We're entering a new era where we're going to have useful drug therapies for treating alcoholism and retaining abstinence," said Dr. Enoch Gordis, director of the National Institute on Alcohol Abuse and Alcoholism.

The dopamine circuit is the heart of the reward system. All roads lead to it. The other circuits—GABA, NMDA, serotonin, and opioids— produce pleasure, from euphoria to contentment, by increasing dopamine.

Although alcohol works on all the reward circuits, its final pathway is through dopamine. Cocaine and amphetamines are strongly addicting because they act directly on dopamine, sharply increasing the neurotransmitter to provide a momentary rush of intense pleasure.

When a person becomes an alcoholic, or addicted to another drug, his or her reward system is changed, providing researchers with new insights as to the causes of tolerance and withdrawal symptoms.

The brain always wants to maintain an even balance. So when it gets too much stimulation from alcohol or cocaine, it tries to blunt the effect of these chemicals. The brain does so, in the case of alcohol, by increasing the number of NMDA receptors to compensate for those that are turned off by chronic alcohol ingestion.

Since NMDA receptors, or "go" switches, are essential for learning, the brain builds more of them in an alcoholic milieu so as to retain its ability to form new memories.

This leads to tolerance. As the brain fights back by building more receptors, alcohol's effect on the brain diminishes and a person has to drink more to get the same high.

The groundwork is then set for withdrawal symptoms. The trouble starts when alcohol consumption ceases. Now the brain has far more NMDA receptors than it normally needs and they are not being subdued by alcohol.

They all begin to clamor with activity. Brain cells become overexcited, causing a spectrum of withdrawal symptoms from feeling on edge to delirium tremens, to seizures. In extreme cases of excitement, brain cells can die.

Scientists no longer believe the old myth that one drink kills thousands of brain cells. Instead it now seems clear that cell death is a byproduct of cells becoming overly excited between bouts of chronic heavy drinking and abstinence.

Fundamentally, all of the receptors that are affected by alcohol go through the same compensatory process as the NMDA receptors.

GABA receptors, which act to stop cells from firing, have the reverse experience of NMDA.

Since alcohol causes them to fire excessively, brain cells try to regain their normal balance by eliminating many GABA receptors. Consequently, when drinking is stopped, there are too few GABA receptors to keep cells in check and prevent them from firing when they're not supposed to. Once again, the effect is to have brain cells become overactive.

"The GABA system would be down-regulated when drinking is stopped," said Grant. "The GABA receptor is implicated in keeping anxiety under control, and a lot of the hangover state is associated with increased anxiousness because GABA temporarily can't do its job."

While alcohol acts in the adult brain to turn receptor systems on or off, in the brain of a developing fetus it can be devastating.

Large amounts of alcohol consumed during pregnancy bathe the developing fetus, affecting almost all its tissues. In the brain, the hippocampus, involved in learning and memory, and the cerebellum, which regulates motor movement, are hit particularly hard by alcohol. Often alcohol's destructive work is carried out through the same receptors that are affected in adults.

One out of every 800 to 1,500 infants born in this country has fetal alcohol syndrome (FAS), which is marked by malformed facial and body features, and mental retardation. FAS is the biggest known preventable cause of mental retardation in the United States.

FAS represents the most severe cases, caused by women who drink heavily during pregnancy.

But for every FAS case, there are many more babies born with subtle developmental or neurological problems that are not reflected in physical appearance, said Dr. Enoch Gordis.

Some of these subtle changes are being uncovered by Nancy Day of the University of Pittsburgh's Western Psychiatric Institute. She has been following 800 children from birth to age ten who were exposed to moderate levels of alcohol—less than one drink a day—during their mother's pregnancy.

She found that these children were significantly smaller and had IQs that were three to four points below average.

"A woman who has a drink once or twice during her pregnancy is obviously not doing any harm to her child," Day said. "But the best advice is to play it safe—no drinks."

Researchers fear that alcohol consumed during pregnancy may leave another dreadful legacy in the brains of the newborns—an enhanced craving for alcohol.

Unlike fetal brains, adult brains can recover. When alcohol is no longer in the brain, receptors eventually readjust to normal levels, unless permanent damage has already occurred. A number of drugs are being studied for their ability to ease or prevent withdrawal symptoms while the brain returns to normal function.

Researchers are also testing drugs that block alcohol's effects on each of the five receptor systems—GABA, NMDA, serotonin, opioid, and dopamine—with early promising results.

"Real breakthroughs are occurring and they are accelerating," said Scripps Research Institute's Koob. "We've got the tools now and we're going to find out where alcohol works to produce its effects and then we're going to find better ways to prevent dependence."

Part Three HOW THE BRAIN FIXES ITSELF

12 Repair and Renewal

After decades of helplessly watching Alzheimer's, Huntington's, and other diseases ravage the brain, medical science is striking back with a potent weapon—the brain's newly discovered power to heal itself.

Enlightened by the insights of molecular biology, scientists now know that Mother Nature equips the brain with a kind of fountain of youth—hormones and other chemicals that nurture and sustain brain cells. But when the fountain begins to dry up, as it sometimes does with age and some mental disorders, brain cells wither and die.

Memory loss, Alzheimer's, Parkinson's, Huntington's, and other degenerative diseases are now believed to be the biological desert created when these rejuvenating chemicals vanish. If we can measure when our brain-nurturing chemicals start to decline and restore them to youthful levels, we may be able to cure or prevent many of the things that go wrong with the brain.

To reach that goal, researchers are uncovering the sources of the brain's fountain of youth. Among the most significant new findings:

- Hormones—estrogen, progesterone, testosterone, and growth hormone—play key roles in maintaining many types of brain cells. Some of these hormones, which may become the first effective drugs to prevent Alzheimer's disease and memory loss, already have produced promising preliminary results.

- Drugs that improve learning and memory in animals are being tested in people, and one of them, a drug used to treat stroke patients, has succeeded in improving human memory.

- Brain chemicals called neurotrophic factors keep cells healthy and communicating with each other. When these factors diminish or disappear, the brain cells they nourish shrivel up. American scientists are gearing up to test one of them, nerve growth factor, to determine whether it can stop the destruction caused by Alzheimer's disease. Other neurotrophic factors may make it possible to grow new brain cells to replace missing ones.

- Brain cells, like muscles, need exercise in the form of education or other stimulating experiences to stay healthy. Researchers have found that the brain's great capacity to physically change and become more powerful with experience—once thought to be limited to childhood—remains strong throughout life.

Stimulating experiences can turn up the flow from the brain's fountain of youth. Education, for example, is a powerful protector against Alzheimer's disease, presumably because it spurs the construction of more connections between brain cells, which can then better withstand the destructive forces of the disease.

In a study of 10,000 Canadian residents, researchers found that education made a big difference in determining who gets Alzheimer's disease and who doesn't. In the study, 7.8 percent of the women over age sixty-five had Alzheimer's while only 5.5 percent of the men had the disease. But when University of Ottawa scientists corrected these rates for education and age, they found that lack of schooling accounted for the higher rate among women.

The implication was that the women in the study, who averaged considerably less education than the men, were more defenseless against brain-cell death and thus more susceptible to Alzheimer's.

As some of the hormones and growth factors that make up the fountain of youth begin to dry up, the brain, like a machine deprived of oil, sputters, clanks, and grinds to a halt. And, as with the Tin

Man in *The Wizard of Oz,* it can be started up again when the missing hormones or growth factors are restored.

"We have to think of the brain in terms of a system in balance," said neurobiochemist Dr. Eugene Roberts of the City of Hope's Beckman Research Institute in Duarte, California.

"When a part of the system changes—like a decline in estrogen, growth factors, or other hormones—the balance is upset and that's when the trouble starts," he said. "What we see as memory loss or Alzheimer's disease can be caused by breakdowns in different parts of the system that end up producing the same symptoms."

Realizing that different causes—hormonal, genetic, environmental—can produce common problems is an important new way of looking at what can go wrong with the brain and how it can be fixed.

Alzheimer's disease, for instance, is now thought to have a number of different causes, even though the result—loss of memory—is the same. As each cause is identified, it opens the door to new treatments for Alzheimer's, including estrogen replacement, fetal tissue transplants, replacement of declining growth factors, and mental exercises. A practical way to administer some of these treatments has so far eluded researchers, but they are optimistic about their prospects.

"We are finally beginning to find the rules that must be followed to keep the brain working in balance," said Roberts, who has shown that he can improve memory in aging animals with a brain hormone called DHEA. Roberts is now testing the precursor of DHEA, a basic hormone called pregnenolone, in human beings to determine whether it improves their memories.

Foremost among the brain's operating rules, which have come to light only in the last five to ten years, is that it constantly rewires itself, even in old age, changing physically and chemically in response to an individual's experiences.

Other rules are being discovered, including those that govern how hormones and neurotrophic factors help brain cells lay down memory traces, and how a shortage of these chemicals causes brain cells to stumble into an Alzheimer's wasteland.

"Anything you learned two years ago is already old information," said neuropsychologist Jeri Janowsky of the Oregon Health Sciences University in Portland. "Neuroscience . . . is exploding."

Fortunately, most brains work well throughout life, diminishing only slightly in memory power with age. For many who do forget easily, the problem may simply be a matter of disuse—in effect, they allow their brains to rust.

For those who have more fundamental problems with their memories—the chemical and electrical circuits that make people who they are—researchers are learning which chemicals in the fountain of youth need to be turned up.

One of their biggest surprises is the discovery that Mother Nature endowed one of her favorite hormones, estrogen, with an ability to nourish many types of crucial brain cells.

Estrogen was once thought to be solely a female sex hormone involved in reproduction. But the hormone, a small, almost indestructible molecule with a biological passport to enter most cells, is turning out to be an important rejuvenator of female and male brains.

"People, and this is true for most doctors, are not aware of the fact that the brain is a major target of estrogen," said cell biologist Dominique Toran-Allerand of Columbia University.

Because of estrogen's ability to carry all sorts of biological messages, it has played a dominant role throughout evolution as a communications superhighway between the brain and the rest of the body in most living things.

That becomes most obvious in women just before menopause, usually around age fifty, when the ovaries drastically decrease their production of estrogen. As estrogen dries up, the brain panics, sending out potentially dangerous false alarms that the body is pregnant, said Dr. Frederick Naftolin, chief of obstetrics and gynecology at the Yale University School of Medicine.

The body begins to extract calcium from bones to begin milk production, but in the postmenopausal years, the loss of calcium merely makes the bones weak and osteoporotic. Fats are liberated into the bloodstream to make the lipid part of milk, but with nowhere to go they are deposited in arteries, accelerating the risk of heart disease.

Sleep becomes light because the brain thinks a woman should be awakened easily if her child stirs, and her body becomes flushed, an evolutionary response designed to keep a cuddled newborn warm in a cave or other poorly heated environment, Naftolin said.

All of these can be delayed or prevented by replacing the missing estrogen.

The brain itself appears to suffer when estrogen levels drop; the risk of Alzheimer's disease, for instance, increases in women after menopause. According to some estimates, postmenopausal women have 5 to 10 percent more Alzheimer's than men, whose estrogen levels do not plummet as rapidly. Several studies suggest that women who take estrogen supplements after menopause can significantly reduce their risk of Alzheimer's.

The male brain also is bathed in estrogen. Naftolin discovered that the female sex hormone is produced by an enzyme that cleaves the male sex hormone testosterone into its components, one being estrogen.

Columbia's Toran-Allerand recently showed that estrogen stimulates the production of nerve growth factor, which is essential for the survivability of many types of brain cells, especially those involved in learning and memory. As estrogen declines, so does nerve growth factor.

Estrogen also increases the sprouting of connections between brain cells. Animal studies in Bruce McEwen's neuroscience laboratory at Rockefeller University showed that by increasing estrogen, more of these connections were formed. When estrogen was reduced, the connections broke off and retracted. It is these connections that determine the power of the brain to produce thoughts, learn new things, move body parts, and respond instinctively.

Thirdly, estrogen prevents a decline in acetylcholine, the chemical messenger that orders new memories to be imprinted in various parts of the brain. Alzheimer's patients suffer increasingly severe losses of acetylcholine, which first robs them of their short-term memory and eventually of their long-term memory.

Meharvan Singh, a neuroscientist at the University of Florida in Gainesville, found that when he removed rat ovaries, thereby terminating their estrogen production, the animals had a 50 percent drop in the enzyme that makes acetylcholine. When he gave the animals estrogen supplements, the key enzyme remained at normal levels and the rats were better able to solve problems, such as avoiding electric shocks, than animals not given the hormone.

"Whatever it is that makes women and men at risk for Alzheimer's disease, if the brain is deprived of a necessary molecule [estrogen], then it may make those neurons [brain cells] more vulnerable to the disease," Toran-Allerand said.

All women do not experience the same rate of estrogen loss. For some women the loss is severe, while other women can make adequate amounts of estrogen in their fat cells. Thin women, for example, are at greater risk for Alzheimer's disease.

"Not everybody needs estrogen replacement," McEwen said. "The entire brain is not going to shrivel up when you take away estrogen. But if you do show estrogen deficiencies, [replacement] might work. This business of individual differences is very powerful and very important."

Barbara Sherwin, a psychologist at McGill University in Montreal, gave estrogen replacements to women whose ovaries were removed during hysterectomies, thus eliminating their major source of estrogen. She found that the brains of the women who got estrogen supplements worked better. The supplemented women could think and remember more effectively than the women not given estrogen. "The machinery is still there and it will continue to run if you keep it well oiled," she said.

Yet, some people resist hormone replacement on the ground that menopause is a natural event. "It may have been natural in 1850 when women lived to be fifty and very few outlived menopause and its consequent problems of osteoporosis, heart disease, and other degenerative disorders," Sherwin said. "Now that North American women live to the average age of seventy-eight, these disorders have become huge problems.

"We've already interfered with life," she continued. "We don't die when we were supposed to because we have antibiotics, coronary bypass operations, kidney dialysis, and other life-saving measures. Now we have to stretch those successes to the brain."

Dr. Stanley Birge, director of the program on aging at Washington University School of Medicine in St. Louis, agrees. "I think we have to recognize that, in fact, estrogen deficiency may not be normal: that it does represent a pathologic state, and that along with osteoporosis, heart disease, and the other problems, we now have to add dementia," he said.

Estrogen replacement therapy can give women the quality of life they had before menopause, Birge said. Among thirteen female Alzheimer's patients over the age of seventy, he found that those who were given estrogen had improved mental function while those who didn't get the hormone continued their gradual slide into mental oblivion.

The results are preliminary but encouraging. Neither the patients, their families, nor the doctors knew who was getting estrogen and who was getting a placebo. But at the end of the eight-month experiment, family members and doctors were able to figure out who was getting estrogen because they could see which patients got better.

Some women shy away from estrogen replacement because of concerns that it may increase their risk of breast cancer. Studies attempting to link estrogen to a breast cancer risk are inconclusive, and some experts believe there is no clear evidence for such a risk.

While there is some evidence for a potential link between estrogen replacement and uterine cancer, many medical authorities consider the risk small. They also note that uterine cancer can be detected early and is curable.

Nobuyoshi Hagino of the University of Texas Health Sciences Center in San Antonio is giving female Alzheimer's patients 0.625 mg of estrogen, the same amount routinely given to women after menopause because it is thought to be safe.

Hagino's study, which involves fifteen women in Japan, found that two thirds of the women improved significantly, and their improvement has continued for up to five years so far.

"They can remember in space and time what happened the day before," he said. "They can communicate with their families and take care of themselves. Patients who wouldn't eat even when food was placed before them in bed, now go to the cafeteria, get their food, eat it, and put their trays on the counter."

Major support for estrogen's potential role against Alzheimer's and other memory disorders came from two long-term studies conducted by Dr. Victor Henderson, professor of neurology, gerontology, and psychology at the University of Southern California.

In the first study of aging and dementia, 253 women were matched for education, age, and symptoms. Of those who had been placed on estrogen therapy to treat postmenopausal problems, 7 percent were eventually diagnosed with Alzheimer's disease, compared with 18 percent of women not taking the hormone.

In the second study, USC researchers studied 8,879 women living in Leisure World, a retirement community in Southern California, between 1981 and 1992. Again, the women were matched for age and other factors. Those who were on estrogen replacement therapy had a 30 to 40 percent reduction in their risk of developing Alzheimer's compared with women not on estrogen.

"I don't want to oversell these results," Henderson said. "We still consider them preliminary. But they suggest that estrogen replacement might be useful for preventing or delaying the onset of this dementia in older women."

As tantalizing as estrogen's role as a brain-saver appears, not all scientists are convinced.

"I think estrogen may have some protective effect but I'm skeptical whether it's as big as everyone thinks," said Dr. Elizabeth Barrett-Connor, head of family and preventive medicine at the University of California at San Diego.

Her study of 800 highly educated women living in Rancho Bernardo in Southern California failed to find any significant memory benefit among those taking estrogen. The women, whose average age was seventy-seven, were basically normal and healthy and were tested for memory impairment associated with aging.

What scientists are counting on to clear up the picture is a large, multicenter, carefully controlled random study in which Alzheimer's patients are given either estrogen or an inactive placebo. The study, which will take several years, is sponsored by the National Institute on Aging.

Helping scientists identify people at risk of developing Alzheimer's is the discovery of a gene linked to the disease. Dr. Allen D. Roses of Duke University found that people who have two copies of a gene for apoE-4, a cholesterol-ferrying molecule, develop Alzheimer's before age seventy. One copy of the same gene is inherited from each parent.

People who don't have any copy of the apoE-4 gene usually don't develop Alzheimer's until after age eighty-five while those with different forms of the gene, apoE-2 or apoE-3, usually don't develop the disease until past the age of ninety. The apoE-2 variety of the gene appears to provide significant protection against Alzheimer's. Scientists are trying to figure out what it is about apoE-2 that is protective in hopes that it can be used to delay the onset of the disease in people who are at risk.

Scientists also are racing to test other chemicals from the brain's fountain of youth. The National Institutes of Health is supporting studies at eight centers to find out whether, by restoring to youthful levels three hormones that decline with age—estrogen, testosterone, and growth hormone—aging can be retarded.

The answer, according to early results, appears to be affirmative, supporting the pioneering work of the late Dr. Daniel Rudman of the Medical College of Wisconsin. Rudman, who reported in 1989 the first evidence that growth hormone increased muscle mass and decreased fat deposits in elderly men, described the age-related decline of testosterone and growth hormone as the "male menopause."

And, like estrogen, testosterone and growth hormone supplements appear to make the brain run better.

"There seems to be a beneficial effect of testosterone supplementation on spacial cognition, which is the ability to interpret one's surroundings in a three-dimensional framework," said Dr. Eric S. Orwoll, chief of endocrinology and metabolism at the Portland Veterans Administration Medical Center.

Orwoll, who is also at the Oregon Health Sciences University in Portland, is treating 100 elderly men with testosterone supplements. So far no side effects have been found.

Early results from other research show that growth hormone, which until recently was thought to be active only in childhood, has a profound effect on brain operations throughout life.

More than 100 adults who have growth hormone deficiencies because of pituitary tumors are being given supplements of the hormone in a study by Dr. Bengt-Ake Bengtsson, chief of endocrinology at the Sahlgrenska Hospital in Göteborg, Sweden.

"The most important thing is the effect of growth hormone on the brain," he said, "improving energy, vitality, sleep patterns, memory, concentration, all those things."

Growth hormone, Bengtsson found, not only acts to make the body physically stronger and spryer, it also alters the chemistry of the brain. One of the changes he found is an increase in beta endorphin, a "good feeling" brain chemical that is linked to memory impairment and depression when it is low.

The only side effect seen so far from growth hormone supplements is fluid retention, which can be easily eliminated by lowering the dose of the hormone.

As scientists discover that there is more than one way to damage or destroy brain cells, they are learning that there is more than one way to save them.

One simple but potentially important way to rescue brain cells from Alzheimer's disease was discovered recently by scientists who wondered why rheumatoid arthritis patients who take anti-inflammatory drugs seldom are struck by the memory-robbing disorder.

Their curiosity led to the theory that Alzheimer's might be linked to a low-grade inflammation in the brain. The inflammation, they reasoned, would slowly attract certain proteins, called membrane attack complex, from the body's immune defense system. The proteins have the nasty habit of chewing up healthy cells, which is what Alzheimer's disease does to the brain.

Because anti-inflammatory drugs given to arthritis sufferers also seep into the brain, they would act on any low-grade inflammation there, keeping the destructive proteins at bay and lowering the patients' risk of Alzheimer's, the scientists theorized.

That's what seemed to be happening when Dr. John Breitner, chief of geriatric psychiatry at the Duke University Medical Center in Durham, North Carolina, looked at fifty sets of elderly identical and nonidentical twins.

Identical twins share 100 percent of their genes while nonidentical twins share 50 percent. Alzheimer's is believed to have a genetic component that makes some people more vulnerable to the still-mysterious causes of the disease.

When Breitner reviewed Alzheimer's cases in which one twin was affected by the disease, or developed it earlier than the other, he found that the protected twin had been taking anti-inflammatory drugs.

In both identical and nonidentical twins, the twin taking anti-inflammatory drugs was four times less likely to develop Alzheimer's or to develop it many years later. An identical twin taking anti-inflammatories had ten times the protection against Alzheimer's as his or her counterpart.

Following up on this line of research, neuroscientist Joseph Rogers, director of the Sun Health Research Institute in Sun City, Arizona, gave the anti-inflammatory drug indomethacin to fourteen Alzheimer's patients and a placebo to fourteen other patients.

Alzheimer's disease worsens over time. But patients given indomethacin not only did not get worse, they improved slightly. Patients given a placebo, on the other hand, continued their mental decline.

"Thc basic research data strongly suggest that inflammation is not only present in the brain [of Alzheimer's patients] but it's doing damage," Rogers said.

Leon Thal, a neuroscientist at the University of California at San Diego, who heads an Alzheimer's clinical trial consortium for the National Institute on Aging, said experiments are under way to determine whether anti-inflammatory agents can delay or prevent Alzheimer's.

"We're moving into a new era where, for the first time, we are beginning to test agents that may slow the rate of decline of memory loss," he said. "That's a huge difference. We have the opportunity to increase the useful life span meaningfully, in a rather dramatic way."

13 Keys to Better Memory

Memory, a trait that makes humans unique among the earth's creatures, has long eluded efforts to understand why it can slowly slip away with age. Now, seemingly overnight, scientists are unlocking its biochemical secrets and enhancing memory in ways that promise to speed learning, halt aging's forgetfulness, and arrest Alzheimer's mental thievery.

Researchers at Northwestern University Medical School in Chicago already are using a drug to boost memory in normal older adults by 50 percent as measured on a test, and scientists from centers around the country are devising other memory pep pills.

Researchers have picked and poked at the brain's 100 billion cells for more than a century. But only in the last decade, especially the last few years, have they begun to learn how the cells work, individually and together.

"We're now saying that we really do think we understand the memory machine," said neuroscientist Gary Lynch of the University of California at Irvine. "But most important is that we have achieved the goal of being able to reach in and tap directly into the machine to improve memory." Lynch has custom-designed chemicals that increase memory in animals.

Among the parts of the machine scientists now can change is the brain's communication system—"signal boxes" sitting on cell surfaces that send and receive chemical messages. And they can regulate "doors" on the cell surface that let electrically charged particles in and out, providing a cell with energy.

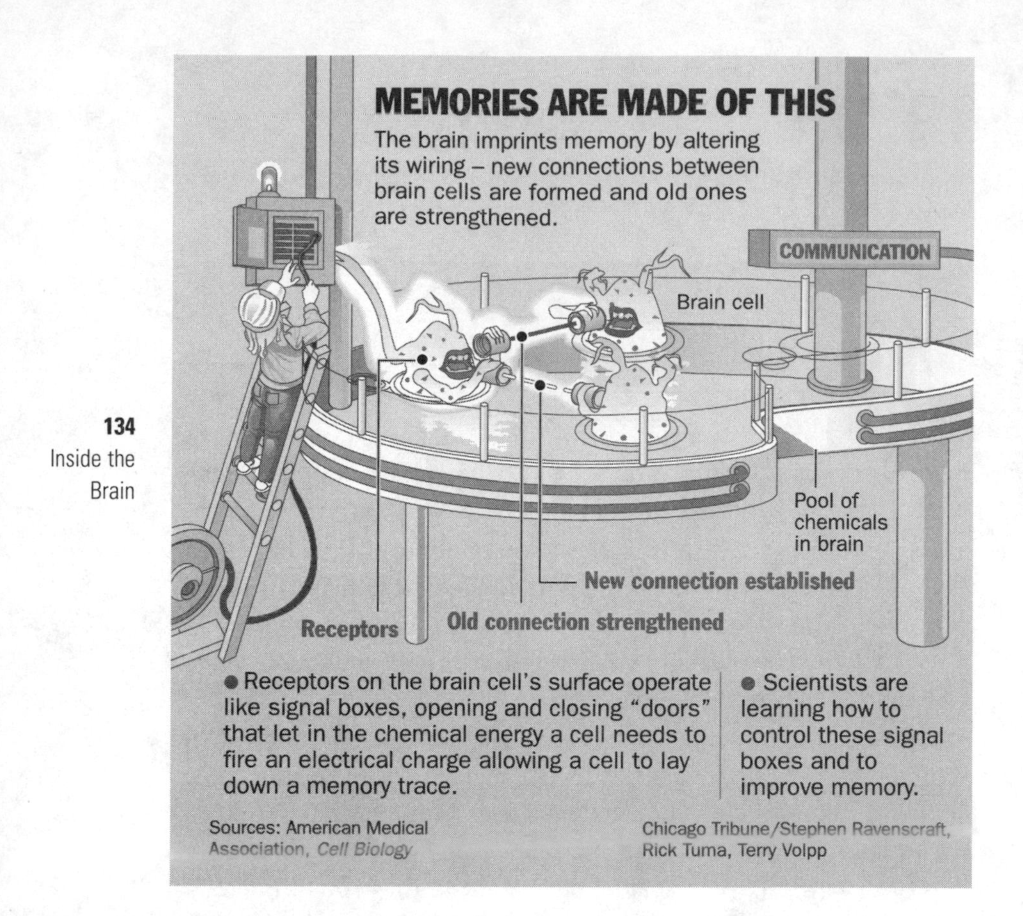

Inside the cell, scientists can regulate chemicals that relay instructions from the outside to a variety of internal workstations, including the cell nucleus, where genes can be turned on and off.

And researchers can direct the chemicals that send tiny electric bolts thundering from cell wall to cell wall to acknowledge that a message from another cell has been received, understood, and acted upon.

The result of all this electrochemical activity is a physical change in the brain—its wiring is altered. Branchlike connections, or synapses, that communicate with other cells squiggle and change as they lay down new memory traces. As trillions of such connections change, they become the physical embodiment of the brain's great capacity to think.

With their newly found ability to influence these physical changes, researchers are confident that memory can be improved.

"The probability is high that there will be at least a moderately effective memory drug available in the not-too-distant future," said James L. McGaugh, director of the Center for the Neurobiology of Learning and Memory at the University of California at Irvine. McGaugh has developed six memory-enhancing agents, including one he hopes to test soon in humans.

Leading the rush to develop memory aids is Northwestern's John Disterhoft with a drug called nimodipine. Nimodipine acts like a gasket to tighten up some of the "doors" on brain cells. As cells age, the doors begin to loosen and allow calcium to slowly leak in and interfere with the chemistry of learning.

Normally, calcium bursts in through the doors to help carry an electric charge from one end to the other. But if the doors leak, calcium molecules dribble in, and their capacity to ignite an electric charge is diminished: There is no sudden surge of current, the message does not get through, and learning does not take place.

This type of memory decline easily can be seen in older people who undergo the eye-blink test, an accurate way to measure memory that does not require verbal response or conscious physical movements.

In the eye-blink experiment, messages enter the brain through an ear and an eye. A tone sounds, followed a half-second later by a gentle puff to the eye. The eye blinks, a seemingly simple response. But something very complex has been set in motion deep inside the brain by the tone-puff sequence. The process of encoding memories, ranging from those as simple as the eye-blink to Einstcin's theory of relativity, starts to take place in an evolutionarily ancient part of the brain, the hippocampus.

The hippocampus is the Grand Central Station of memory. It dispatches arriving trains of thoughts to either short commuter runs that are quickly forgotten—phone numbers, names of party guests—or to more permanent destinations in the brain where important things like your home address, spouse's name, and the Second Law of Thermodynamics are stored.

When the tone-puff messages arrive at the hippocampus, it says,

"Wait a minute. Every time I hear the sound I get a puff. If I blink when I hear the sound, I can avoid that annoying puff to the eye."

This type of memory encoding is called associative learning—the kind taught in the classroom and the kind that people use to stoke their imagination and creativity.

What the hippocampus is doing is changing incoming messages from short-term commuters to long-distance travelers when it decides it needs to save something like blinking at the sound of the tone.

The brain encodes the learning, and young brains do it more efficiently than older ones. Out of fifty tone-puff trials, the average young adult in his or her mid-twenties automatically will respond thirty-five times, blinking after the tone and before the puff.

That is pretty successful learning. But people between the ages of sixty and seventy-five, who have leaky doors on their brain cells, will not learn as well. On average they blink only twenty times after fifty tones. Failure to learn means they get more puffs to the eye.

But when older adults were given nimodipine tablets three times a day for three months, they increased their blinks from twenty to thirty, a 50 percent leap in their ability to learn this task over those given placebos instead of nimodipine.

"I don't want to oversell this," Disterhoft said. "I don't think it's a cure for aging or anything. But the effects are quite impressive."

So far nimodipine has had few side effects and the drug already has been approved for treating stroke patients.

While Disterhoft tightens cell doors to improve memory, the University of California's Lynch tinkers with the cell's communications system—its "signal boxes."

Scientists recently discovered a key signal box, called the NMDA receptor, which makes the grand decision to fire an electric charge through the cell, thereby blazing a memory path.

The NMDA signal box sits on the cell surface, awaiting a message from glutamate, a neurotransmitter that races between brain cells like a bicycle messenger yelling, "Pay attention, there's something coming that you might want to learn."

When the glutamate messenger lands on the NMDA receptor, it opens certain doors on the cell surface, letting in energy-carrying sodium particles. These positively charged particles act like batteries to build up the cell's electric charge. When a critical point is reached, a biochemical "spark" is ignited and the cell's connections to other cells are changed—a memory pathway has been forged.

People often block this process with alcohol. Studies show that the reason people can't remember things after heavy drinking is because alcohol molecules clog NMDA receptors. Glutamate messengers bounce off and no signal gets sent.

Knowing how the NMDA receptor works, Lynch and Gary Rogers of the University of California at Santa Barbara devised a chemical that increases the brain's level of attention. The chemical, a member of a family of compounds called ampakines, fits unobtrusively into the receptor.

Normally a glutamate molecule plugs into a receptor, delivers its message, and departs in one-thousandth of a second. But the ampakines, like a wad of gum stuck to the sole of a shoe, make glutamate stick around twice as long.

As a result, the cell doors remain open twice as long and more sodium particles pour in. The NMDA receptor, alerted that something important is happening that should be remembered, fires a charge.

People can do the same thing naturally when they decide to pay greater attention to something they want to learn. Ampakines increase the attention level once they have been aroused, so they make learning easier, Lynch said. "For the first time we are in the position of being able to say that we designed a drug to hit a specific part of the memory system and that it does improve memory."

Healthy young rats given ampakines learned tasks in half the normal time. No side effects have been noted, and Cortex, a new drug company devoted to developing memory-enhancing agents, plans to start human trials with the ampakines, said Lynch, who is a company co-founder. "We're talking about a pill that you would take, and two minutes later and for the next several hours, information that you're trying to encode in your brain is going to be encoded better."

Can memory enhancers overload the brain with too many memories? Not likely. A neural network model of only 1,000 brain cells, each having a couple hundred connections, was able to learn 10,000 words randomly selected from a dictionary without filling up the system.

"The brain turns out to be a very efficient machine for encoding tons of information," Lynch said. "You'd be amazed at how few connections you need to alter to encode a memory. It's minuscule."

A brain cell fires an electric charge so it can rewire itself as a new memory is formed, and so it can send the new information to other cells.

Northwestern's Aryeh Routtenberg believes he has figured out how those changes are made, at least in a broad sense. In 1984, Routtenberg discovered an important chemical inside brain cells called PKC, which forms short-term and long-term memory.

Once a cell is alerted to the fact that it needs to form a memory, a second chemical messenger notifies PKC. PKC then runs over to another protein, F1, and pitches a phosphate molecule at it. The phosphate wakes up F1 and it trundles down the branches of the cell to change the synaptic connections.

These are temporary changes, lasting long enough for a person to dial a restaurant number after looking it up in a telephone book.

When a more permanent memory is required, PKC zips to the nucleus and launches a genetic process that results in the production of more F1. With an army of F1 engineers busily rearranging the architecture of the brain-cell connections, more permanent memories can be formed.

Yuri Geinisman of Northwestern's Institute for Neuroscience showed how synapses—the communication points between brain cells—are altered physically to become stronger with learning.

Geinisman found that learning changes the surfaces of synapses from flat to bumpy. The bumps form compartments that store an increasing amount of information. With disuse, these compartments fade and the surfaces flatten out, erasing the memories that were once there. Besides strengthening synapses, brain cells also can make new connections to enhance learning.

Because the second messenger that activates PKC is made up of a type of fat commonly found in brain-cell membranes, Routtenberg fed various oils to mice and found that one, corn oil, stimulates PKC production. The animals also learned about 15 to 20 percent faster than control mice.

Based on Routtenberg's research, an Italian drug company developed a chemical that mimics corn oil's effects on PKC. Preliminary tests in Alzheimer's patients showed a slight improvement in memory.

Just as the brain has mechanisms for stitching memories together, it has mechanisms to unstitch them, and one of these may lead to ways to delay or prevent the memory loss from Alzheimer's.

Beta amyloid is a neurotransmitter that is overproduced in the brains of Alzheimer's patients, and scientists have puzzled over what role it plays in the disease.

Dr. John E. Morley, chief of geriatric medicine at the St. Louis University School of Medicine, found in animal studies that beta amyloid appears to disassemble memories that are no longer needed.

"Its normal role within the brain would be to help us forget things we didn't want to remember or to erase memories that have never been useful," he said. The overabundance of amyloid also may cause the low-grade inflammation that is suspected as an agent in Alzheimer's.

But how does it destroy memory? If Alzheimer's patients are making too much beta amyloid, perhaps their memories are being dissolved before they can be solidified. Might the same be true for normal, healthy older people who make more beta amyloid as they age?

To find out, Morley made a series of "blanks," proteins that occupied the same brain-cell "signal boxes" as beta amyloid but did not transmit any messages. Giving the blanks to aging mice, whose memories rapidly fail, he found that beta amyloid was barred from plugging into the receptors. The memories of the animals got better.

Amyloid blockers are not yet a feasible treatment for humans because they cannot cross the blood-brain barrier, the brain's defense against potentially harmful compounds. Similar chemicals need to be developed that can cross the barrier and shut off the beta amyloid receptor.

Other scientists are exploring different ways to save memory. John Carney of the University of Kentucky and Robert Floyd of the Oklahoma Medical Research Foundation hope to slow memory loss with drugs that prevent damage to brain cells.

Using a super-antioxidant called PBN, they have been able to improve memory in aging animals. Antioxidants reduce damage to genes and cell structures from free radicals, which are particles produced as a result of normal cellular chemistry. In some cases, too many free radicals are formed and they can damage brain cells. Excessive beta amyloid production, for instance, may cause its inflammatory damage through free radicals.

PBN quenches the free-radical reaction, thereby giving cells time to repair routine damage, Carney said. "There's a critical point in aging where free-radical damage outstrips the ability of cells to repair themselves."

The new science of molecular biology is spawning unexpected approaches to fine-tuning the brain. Joseph Moskal, research director of the Chicago Institute of Neurosurgery and Neuroresearch at Columbus Hospital, is using the new science to develop "keys" that fit into the brain's signal boxes.

One of his keys, which fits a receptor called glycine, helps trigger brain cells to fire electric charges and speed learning. Used in animals, the key increased their learning twofold as measured by a memory test.

"It sounds like science fiction, but with molecular biology we can . . . hopefully generate all sorts of interesting novel compounds that enhance cognition," said Moskal, co-founder of a biotech company called Neurotherapeutics.

The company, established with a small-business innovative research grant from the National Institute of Mental Health, hopes to test some of its memory-enhancing keys in humans within a couple of years.

While researchers have developed dozens of drugs that improve memory in animals, it is far more difficult to make a memory pill that is effective and absolutely safe for humans.

An example is tacrine, the only memory booster that has been approved for Alzheimer's disease. Tacrine increases levels of an important chemical messenger, acetylcholine, which encourages crosstalk among brain cells. Acetylcholine declines as Alzheimer's disease progresses, cutting communications and blocking would-be memories.

Tacrine slightly improves memory, attention, reasoning, and language in some Alzheimer's patients but it can cause serious side effects, and the disease process continues to take its toll.

So far, drug companies are much better at making drugs that erase memory. A number of medications on the market, such as Valium and other antianxiety compounds, and sleeping pills such as Halcion, diminish memory. But if some drugs easily can turn off memory, researchers believe, others can turn it on.

Nondrug ways to improve memory are also being explored. Music, for instance, besides being enjoyable appears to be able to increase brain power, possibly by exercising the same circuits employed in memory formation.

Psychologist Frances Rauscher of the University of California at Irvine first showed that college students listening to ten minutes of Mozart's Piano Sonata, K 448, increased their spatial IQ scores. Spatial intelligence, the ability to accurately form mental images of physical objects and to be able to recognize variations in their shapes and positions, is important for higher brain function, especially the type of reasoning used in physics, mathematics, and engineering.

Toddlers could also benefit from music lessons, Rauscher found. Nineteen preschool children taking eight months of music lessons performed far better on spatial reasoning tests than similar children who were not given lessons.

Classical music or jazz appears to invigorate the same brain areas used for spatial reasoning. Atonal music, such as that of Philip Glass, or highly rhythmic dance pieces, that clang instead of flow, failed to improve spatial scores.

As their brain-adjusting tool kit enlarges, many scientists believe that the power to routinely improve memory is just around the corner.

"We are close," said St. Louis University's Morley. "The excitement is there. But the investment needed to make it really happen will have to be on the order of the investment going into AIDS research.

"Given that most of us are going to get Alzheimer's disease instead of AIDS, the investment is well worth it."

14 Chemicals That Fix the Brain

The discovery of how the brain heals itself began, as discoveries often do, with a question: Why do children who suffer brain damage often recover fully, while adults with the same kind of damage are permanently incapacitated?

University of Wisconsin neurobiologist Ronald Kalil was among those who pursued the question. His studies in young cats showed that entire networks of brain cells could be rerouted around damaged areas. Young animals whose primary vision centers were destroyed could still learn to see normally, he found, because cells in another part of their brains took up the job of processing vision. Yet adult animals suffering the same destruction had no such luck. What was the difference?

Kalil finally determined that young animal brains are awash in chemicals called growth factors, while adult brains have far lower levels. He surmised that the abundance of growth factors helps the new brains organize themselves; when damage occurs, the growth factors simply start over and rebuild damaged networks.

Adults have fewer growth factors because their brains, although they constantly undergo changes, are, for the most part, completed.

All of which led to another question: Would adding extra growth factors prevent permanent damage in adult brains?

Soaking tiny sponges with a variety of growth factors, Kalil placed them inside newly damaged brain areas of adult cats. He and his colleagues found that these adult brains acted more like infant brains: Instead of suffering permanent damage, the adult brains repaired themselves.

"Will we have things on the shelf that will fix the brain? I think so," Kalil said. "Breakthroughs are coming so rapidly. Brain cells that are in trouble can be kept alive and they can be induced to form new connections."

This ability of the brain to rewire itself, grow new parts for damaged cells, and even make new cells—its "plasticity," in scientific jargon—was thought to be impossible only a few years ago. Brain cells, medical students were taught, were hard-wired like so many computer transistors. Once they burned out, that was the end.

Brain cells certainly couldn't sprout new communications lines to take over the jobs of nonfunctioning cells, it was said. Nor could they regenerate themselves after being hurt. And they absolutely could not divide to replenish the brain with new cells.

All those "truths" are being tossed out as brain research undergoes a revolution fueled by molecular biology's remarkable ability to reveal the secrets of cells. Scientists now can hunt down and copy genes that govern cell reassembly and harness them for use in repairing damaged brains.

The power of these tools was stunningly demonstrated with the discovery of a gene called NeuroD, which plays a key role in the embryonic development of the brain and nervous system. So potent is this gene that when inserted into cells that normally would never become brain cells, it transforms them into brain cells.

According to Dr. Jacqueline Lee of the University of Colorado Cancer Center in Denver, who headed the research team while she was at the Fred Hutchinson Cancer Center in Seattle, NeuroD not only opens the door to studying how brain cells are made, but it may make it possible to make new brain cells to replace those destroyed by Alzheimer's and other disorders.

"This organ that seemed so inaccessible, that seemed as if it couldn't be repaired just a few short years ago, now appears to be monumentally plastic, and we are beginning to take advantage of its healing powers," says Dr. Ira Black, chief of neuroscience and cell biology at the Robert Wood Johnson Medical School in Piscataway, New Jersey.

The key to the brain's plasticity is a newly discovered family of chemicals, including nerve growth factor, that keep brain cells alive.

They are called neurotrophic factors, from the Greek "neuro" for brain and "troph" for nourish.

They have names like brain-derived growth factor, glial-derived growth factor, neurotrophic-3 and ciliary neurotrophic factor—clumsy titles that belie their life-saving attributes.

Neurotrophic factors are like super-nannies to brain cells. They are there when the cells are first born, making sure they are nourished, grow, and make the right connections. They are there throughout the life of the cells, guarding their health and repairing damage. And they are there to insure that the brain cells do the jobs they were created for, to learn and remember.

When neurotrophic factors decline or disappear, brain cells quickly fall down on the job, shrink, and eventually die. This happens because the factors are no longer there to protect cells from being chopped up by free radicals, molecular piranhas created by normal body chemistry. Nor are the factors there to turn genes on and off to maintain the cells' communications lines or make sure the cells are sending out their "I'm okay, you're okay" messages to other cells.

The discovery of neurotrophic factors has opened the door to new generations of therapeutic agents that can revive sputtering brain cells, repair those that are damaged, rescue dying ones, and generate new cells.

"They hold the promise that we can correct neurologic problems that in the past were thought to be intractable," said Dr. Eugene Major, chief of molecular medicine and neuroscience at the National Institutes of Health's Neurological Institute.

"They hold the additional promise that we might be able to maintain neurologic function, which we thought always diminished with age," he said. "And, if we really want to be hopeful, we might even be able to provide a higher quality of life."

That goal involves nothing less than the treatment and possible cure of such neurological scourges as Alzheimer's disease, Parkinson's, ALS, Huntington's, age-related memory loss, and stroke. The race to develop these new treatments has already started:

- Two biotechnology companies—Genentech and Syntex—along with the National Institute on Aging are undertaking large-scale trials to restore nerve growth factor to the brains of patients with Alzheimer's, a memory-robbing disease affecting 4.5 million Americans. Implantable pumps that bathe the center of the brain with nerve growth factor will be used to determine if NGF can halt the destruction of Alzheimer's and bring back memory.

- Glial-derived growth factor, which was recently discovered by Synergen's Frank Collins, has been shown in laboratory tests to save brain cells that die in Parkinson's patients. Company officials hope to start testing GDNF in patients soon to determine if it can reverse the disease.

Another growth factor, GM1 ganglioside, substantially reduced some of the functional impairments in Parkinson's patients, such as rigidity, tremors, and cognitive problems, according to the preliminary results of a pilot study. The growth factor, which was obtained from cow brains, appears to increase the brain's supply of dopamine.

"Though the precise mechanism is not clear, GM1 ganglioside may rescue damaged dopamine neurons, stimulate them to repair themselves, promote growth of new dopamine terminals on the neurons—and perhaps even stimulate remaining dopamine neurons to produce more than their normal amount of dopamine," said neurobiologist Jay Schneider of the Medical College of Pennsylvania and Hahnemann University. Larger studies are in progress.

That scientists are on the right track to repairing the brain was strikingly demonstrated in another Parkinson's-related study. Researchers reported the first proof that fetal tissue transplants survived, grew, and functioned in the brain of a Parkinson's patient.

The transplant was linked to a significant improvement in the patient's condition, freeing him from the prison of rigidity and immobility, the main symptoms of the disease, and enabling him to enroll in an exercise class. However, the patient died from an unrelated cause. Scientists who examined his brain found that the transplanted tissue had been working.

"For the past fifteen years we've been trying to get to this point, to understand whether fetal transplants can work and how they work," said Dr. Jeffrey Kordower, director of the Research Center for Brain Repair at Rush-Presbyterian–St. Luke's Medical Center in Chicago.

"The grafts survived and they replaced the lost circuitry that is destroyed by the disease," Dr. Kordower said.

The National Institutes of Health is funding further studies to test the benefits of dopamine-producing fetal tissue transplants in the brains of Parkinson's patients.

The disease, which affects 500,000 Americans a year, is marked by a decline in the brain cells that make dopamine, a key neurotransmitter, or chemical messenger. Dopamine loss results in shaking, inability to move, and lack of body control.

- Amyotrophic lateral sclerosis (ALS), also known as Lou Gehrig's disease, strikes 5,000 Americans a year and is characterized by muscle paralysis. The fatal disease appears to be linked to a decrease in neurotrophic factors that protect motor neurons from free-radical damage.

The first success in slowing the devastating progression of ALS has been achieved with insulin-like growth factor. Scientists at eight U.S. and Canadian medical centers found that patients taking the drug, manufactured by Cephalon, Inc., had 25 percent less deterioration over a six-month period than other ALS patients who did not receive the growth factor. IGF-1 supports the survival and regeneration of motor neurons.

More encouraging work is under way for ALS. Studies suggest that injecting motor neurons that activate muscles with another factor, ciliary neurotrophic factor, may also slow progression of the disease. This neurotrophic factor seems to switch on genes that produce chemicals that neutralize free radicals.

- Amgen and Regeneron, biotech companies that have learned how to genetically engineer growth factors, are considering trials to find out if brain-derived neurotrophic factor can stop the destruction of Huntington's disease, a genetically linked deterioration of certain brain cells that affects 30,000 Americans.

Some neurotrophic factors are so powerful they can actually transform some body cells into brain cells. Cells taken from the inner core of adrenal glands, for instance, which sit atop the kidneys, normally make a tiny amount of dopamine.

When mixed with nerve growth factor these adrenal cells change their shape, taking the form and function of brain cells. They sprout connections to nearby neurons and increase their production of dopamine, the neurotransmitter that declines in Parkinson's patients.

Lars Olson of the Karolinska Institute in Stockholm, who conducted the original conversion experiment in the laboratory, is testing adrenal cells in humans with Parkinson's disease.

A small number of cells are removed from a patient's adrenal glands, inserted into the brain through small holes in the skull, and then bathed in nerve growth factor for three weeks. A big advantage of this technique is that it avoids the risk of transplant rejection, since the cells are the patient's own.

"Once they've become nerve cells they have many [connections] that extend out into the host brain and they are able to survive on their own," Olson said. "Now we have nerve cells that make dopamine and that's what we believe causes functional improvement in patients."

Although the improvement is significant, more patients will have to be treated with the converted adrenal cells before conclusions can be made about the technique's safety and long-term effectiveness.

Olson's enthusiasm for nerve growth factor is motivated by memories of two childhood friends who suffered brain damage in motorcycle accidents. Now he sees the possibility of preventing and reversing mental deterioration for the first time, and he is eager to test nerve growth factor wherever it might do some good.

That includes Alzheimer's disease. Olson pioneered the use of minipumps that infused nerve growth factor into the brains of Alzheimer's patients several years ago. His first patient got some memory back and lost it again when nerve growth factor was discontinued as part of the study.

Olson's experiments, along with other evidence—genetically engineered mice that have no gene for nerve growth factor and no memory, and NGF infusions that perk up memory in aged monkeys—have convinced U.S. scientists that it is time to conduct full-scale studies of nerve growth factor in Alzheimer's patients.

"We are encouraged to go forward," said Dr. Zaven Khachaturian, former chief of Alzheimer's research at the National Institute on Aging, where he has set up a consortium of twenty-eight medical centers around the country to test new Alzheimer's medications. "Within the next five years we want to come up with a compound that's going to significantly delay the onset of Alzheimer's symptoms," he said.

Perking up the failing memories of Alzheimer's patients may be possible with genetically engineered skin cells. Dr. Leon Thal of the University of California at San Diego and Lisa Fisher of the Salk Institute showed that they could restore memory in rats by inserting a gene that makes a protein essential for memory recovery into some of the animals' skin cells and then inserting the cells into their brains.

The transplanted gene makes choline acetyltransferase, which turns choline into the neurotransmitter acetylcholine. It is the gradual decline of acetylcholine that causes Alzheimer's patients to lose their memory. Acetylcholine is the oil that makes the memory machine function. When it dries up, the machine freezes.

"This research is encouraging news for Alzheimer's disease treatment because it indicates that delivering a single neurotransmitter may provide memory improvement," Thal said.

While scientists may discover more chemical keys for healing the brain, they still face a formidable obstacle to their use. They have keys to the castle doors, but they still have to get across the moat, the blood-brain barrier.

The barrier's job is to keep potentially harmful chemicals and proteins out of the brain. Neurotrophic factors, which are large proteins produced in the brain, also can't cross the barrier—they stay inside. When scientists genetically engineer neurotrophic factors in the laboratory and then try to add them to the brain through pills or injections, the blood-brain barrier stops them cold.

For now, strategies to bypass the barrier are the equivalent of tunneling under the castle moat: Scientists drill tiny holes through the skull to pump growth factors directly into the brain. They also use the holes to implant fetal brain cells or small packets of skin or muscle cells that have been engineered to contain genes that make growth factors.

While these surgical techniques have produced favorable results, scientists are looking for easier ways to boost growth factors in ailing brains. Surgery is invasive and may not be practical for the millions of people who might benefit from fresh supplies of neurotrophic proteins.

Scientists hope to outsmart the blood-brain barrier just as they outwitted the body's immune-defense system to make organ transplants routinely successful. That was accomplished with drugs that selectively subdue the immune system's ability to reject organs while leaving intact its power to kill germs.

Researchers are developing similar strategies to cross the brain's moat, modern versions of catapults, spies, bridges, and ladders.

One approach involves altering viruses to ferry growth factor genes into brain cells. The newest of these is a "ghost virus" developed by Richard Samulski of the University of North Carolina. Samulski removes all the genes from certain viruses except for two tiny bits of DNA used for making the protein coat that covers the viral genes. In place of the absent genes, Samulski inserts growth factor genes. The protein coat attaches to brain cells and squirts in the neurotrophic genes.

Using the "ghost virus" equipped with a dopamine-producing gene, Rockefeller University neuroscientist Michael Kaplitt has produced significant improvement in rats with brain disorders that mimic Parkinson's disease.

Other researchers are trying to deliver growth factors to the brain by way of the blood. Alkermes, a biotech company in Cambridge, Massachusetts, hooks a growth factor protein onto an antibody that can cross the blood-brain barrier. Biochemical "sentries" guarding the barrier think the antibody is carrying iron, which the brain needs, instead of growth proteins.

Early results in animals show that nerve growth factor gets across the barrier piggybacked on the antibody, and that it revitalizes sick brain cells, said Greg Gerhardt of the University of Colorado.

An even simpler technique is being pursued by Mark Mattson of the University of Kentucky. Mattson is working with a class of small proteins called alkyloids that are produced by bacteria. The alkyloids easily cross the blood-brain barrier and fit into the same receptors on brain cells that are the docking ports for growth factors. Once attached, the alkyloids send the same kind of message to the cell that normal growth factors do.

Perhaps the best way to finesse the blood-brain barrier would be to induce healthy cells to multiply within the brain and replace damaged cells. But until two years ago, it was firmly believed that brain cells don't divide.

That myth was shattered by neuroscientists Samuel Weiss and Brent Reynolds of the University of Calgary. The Canadian researchers were amazed to see the cells divide when they were nourished with epidermal growth factor. EGF even makes adult cells divide.

The dividing cells are called stem cells because they can give birth to a variety of different brain cells, just as stem cells in bone marrow give rise to all types of blood cells.

Understanding how these cells can be manipulated may lead to promoting their production in the brain "and thus allow the brain to repair itself in much the same way skin does," Weiss said.

Anders Bjorkland of Sweden's University of Lund, who paved the way for fetal tissue transplants in humans, is researching brain stem cells in animals to learn if they can replace dying or missing brain cells. So far, it appears they can.

"The ideal would be to make them into any kind of brain cell one wanted," Bjorkland said.

So promising has this line of research become that scientists now believe stem cells can be used to prevent mental retardation caused by inherited metabolic diseases. These genetic disorders affect 1 out of every 1,500 newborns.

The idea would be to insert stem cells into the ventricle spaces in the brains of embryos or newborns affected with these genetic disorders. The healthy stem cells would then multiply and migrate to areas of the brain where they are needed to correct the genetic mistake, thereby insuring the development of a normal brain.

And that's exactly what happened in the brains of mice that had inherited the human equivalent of mucopolysaccharidosis, a disorder in which brain cells do not make an enzyme that breaks down waste products. As a result, waste products build up, damaging the cells and causing mental retardation and early death.

It is a sentence that immutably destroys the lives of thousands of children every year. But it is a sentence that may one day be lifted. In a breathtaking series of experiments, John Wolfe, a medical geneticist at the University of Pennsylvania, Evan Snyder of Harvard and Boston's Children's Hospital, and Penn's Rosanne Taylor prevented retardation from developing in the mice by injecting normal stem cells into their brains.

The stem cells corrected the genetic deficiency. Autopsy studies revealed that the transplanted cells took up residence with the animals' own neurons, moved along with them to different areas of the brain, settled and differentiated into specialized cells appropriate to those brain regions. They became a normal part of the animals' brains, and most critically, they produced the missing enzyme that was able to clear waste products from their brain cells.

Although fetal tissue transplants have shown promise in treating brain disorders such as Parkinson's, many researchers would like to avoid the use of tissue from aborted fetuses. One alternative is to use a person's own skin or muscle cells. These cells can be bioengineered to carry growth factor genes, thus converting them into tiny factories for producing growth factors.

Altering skin cells from monkeys to carry nerve growth factor genes, Fred Gage of the University of California at San Diego then implanted the engineered cells in their brains to prevent the kinds of deterioration seen in Alzheimer's and Parkinson's patients.

Jon Wolff of the University of Wisconsin used genetically engineered muscle cells to do the same thing, and he has devised what is probably the simplest technique to put foreign genes into cells. It's called the "naked DNA" approach and consists of simply squirting bare DNA, the chemical units of genes, into cell cultures. The genes are absorbed into the cells, by some still-unknown process, and start churning out copious amounts of growth factors.

"This could be an outpatient procedure for Parkinson's and Alzheimer's patients," Wolff said. "First you take a muscle biopsy, grow the muscle cells in cultures with growth factor genes, and then a month later inject them into the brain under local anesthetic."

Reflecting on these and other recent advances in brain research, the Karolinska Institute's Olson observed, "I've been working in this field for more than thirty years and it's only been in the last five years that we finally have some reason to be optimistic, thanks to molecular biology and the discovery of all these new growth factors."

15 Keeping the Brain Sharp as We Age

Just as medical researchers discovered a few decades ago that muscles weaken without repeated exercise, neuroscientists are coming to the same conclusion about the brain—it rusts with disuse.

And just as exercise keeps people vigorous into their seventies and eighties, researchers are demonstrating that mental workouts can do the same for the aging brain.

Aging has long been thought to be an irreversible downhill slide into mental befuddlement. But the new research shows that was little more than a self-fulfilling prophecy, usually the result of brain disuse.

Furthermore, people do not lose massive numbers of brain cells each day as they grow older, as was once thought.

"We used to be taught we were losing nerve cells every day, something like a million a day," said Dr. Marilyn Albert, associate professor of psychiatry and neurology at Harvard and director of gerontology research at Massachusetts General Hospital. "We're beginning to get very good evidence that there is not a lot of neuronal [brain cell] loss with age."

When Dr. Monte S. Buchsbaum first trained his PET (positron emission tomography) scanner on the brains of aging volunteers, he expected to see shrinkage in the area where memory is stored—confirming the widely held theory that forgetfulness increases with age as memory cells die off.

"Boy, were we wrong," says Buchsbaum, director of the Neuroscience PET Laboratory at the Mount Sinai School of Medicine in

New York. Punching keys on a computer, he calls up images that compare the brains of people in their twenties to those of people in their seventies and eighties.

He points to the frontal lobe, also called the neocortex, where we think about thinking, where plans are made to call up memories, and decisions are executed as to how to use them in ways that provide new insights, inspiration, imagination, and creativity.

After a twenty-five-year-old is given a half-hour memory test, the frontal lobe glows bright yellow and red. The colorful display is an indication that radioactive sugar, which had been injected into the blood of the volunteers, was consumed as energy by billions of brain cells busily processing memory.

That was expected. The surprise was that the frontal lobe of a seventy-five-year-old shone just as brilliantly after the same memory test.

"The good news is that there isn't much difference between a twenty-five-year-old brain and a seventy-five-year-old brain," said Buchsbaum, who scanned the brains of more than fifty normal volunteers who ranged in age from twenty to eighty-seven.

Findings like these are revolutionizing our understanding of the untapped powers of the brain and they are leading to new ways of thinking of how people can maintain their brains in top shape.

Understanding that muscles need to be exercised had a dramatic effect on frailty in the aged. A generation ago frailty could be seen in people in their sixties. Today frailty is not likely to be seen until people are in their eighties and nineties. Scientists are convinced that the brain can experience that same kind of turnaround.

When there is a severe decline in memory and mental function, it is usually caused by such diseases as Alzheimer's, which affects only 20 percent of the very old. Most of the other decline, the kind that people tend to fret about as they age, is now believed to be caused by a lack of mental exercise.

"The more levels of education you have, the more likely you are to engage in mentally stimulating activities, and that's actually good for your brain," Albert said. Her study of more than 1,000 people from age seventy to eighty showed that four factors seem to determine which oldsters maintain their mental agility:

- Education, which appears to increase the number and strength of connections between brain cells.

- Strenuous activity, which improves blood flow to the brain.

- Lung function, which makes sure the blood is adequately oxygenated.

- The feeling that what you do makes a difference in your life.

"Is mental exercise important for the brain? People used to ask me that years ago and I would say we don't have enough data to say one way or another," Albert said. "I don't say that any more. I tell them that's what the data look like—use it or lose it."

Mental exercise, scientists are finding, causes physical changes in the brain, strengthening connections between brain cells called synapses and actually building new connections.

Such physical changes can occur within seconds, as when we shift attention, or they may take hours or days, as some memories are cast into the biological ingots that last a lifetime.

Education and interesting work protect people against Alzheimer's disease, research shows. The more connections a person has between brain cells, the more resistant he or she is to the onslaught of this memory-robbing disorder.

Dr. Yaakov Stern, a clinical neuropsychologist at Columbia University, found that people who had less than an eighth-grade education had twice the risk of Alzheimer's as those who went beyond the eighth grade. And when people with lower education levels also worked at mentally unstimulating occupations, they had three times the risk of becoming demented.

While the research of Buchsbaum and other scientists shows that the brain remains pretty much intact as it ages, some deterioration does occur. Weak points have been discovered in specific areas of the brain that act as relay switches, passing messages between the memory bank and other parts of the brain.

The relay switches include the basal ganglia, which passes on commands to move muscles; the hippocampus, which decides if memories are to be placed in short-term or long-term storage; and the amygdala, which imprints memories with such emotions as fear and love.

Buchsbaum found that although the memory bank in the brains of older people glows as brilliantly as it does in younger brains, some relay switches, such as the basal ganglia, are a bit dimmer. That means the basal ganglia is less active, almost lazy. As a result, muscle coordination declines, perhaps explaining why many older people tend to slow down.

Other memory problems may occur when different relay switches falter or fail to work.

The new findings may make it possible to identify these dim relay switches and develop techniques to brighten them up again. These could include mental exercises that stimulate specific areas of the brain such as the basal ganglia and the hippocampus, or drugs like neurotrophic factors that can reinvigorate brain cells.

Research indicates that certain exercises can build up specific brain areas, and some scientists are setting up programs to use this new knowledge to help learning-disabled children.

New research also shows that the aging brain retains much of the same capacity as a child's brain to rewire itself. Although adult brains are not quite as good at repair as infant brains, they still do an amazing job.

One of the first recorded examples of such repair occurred in 1909 when psychologist Kenneth Craik did a stupid thing, something children are always told not to do.

Out of scientific curiosity, Craik stared at the sun with one eye, destroying light receptors in the center of the retina. The receptors could no longer send information directly to the brain cells that process vision, so Craik's central vision was lost in that eye.

Ever the scientist, Craik assiduously noted what happened to his sight. Immediately after the damage he couldn't see to aim a rifle. Words were missing from pages and birds vanished from the sky.

But then a strange thing started to happen: Within a month, some of his central vision began to return and by the end of six months he could see normally again.

Only now have scientists figured out what happened to Craik. In doing so they have opened a new door into the brain, shattering the old concept that brain circuits that allow us to see, think, remember, and act were "hard-wired" and never changed.

Rockefeller University scientists who repeated Craik's experiment in animals found that the brain in adulthood and older age can rewire itself. In the case of the animals—and presumably Craik—brain cells that no longer received inputs from the eye's damaged sensory cells made new connections to healthy sensory cells so that they could see again.

Evidence that the brain can repair itself has important implications for repairing damage caused by stroke, for rewiring destructive thought patterns, and for learning new skills.

This ability to change circuits, to change the functional properties of brain cells, is something that originally evolved to help people recover from damage to the brain, theorizes Rockefeller University neurobiologist Charles Gilbert.

But it was such a good adaptive device that the brain used the same mechanism of repair to construct memory. And both memory and repair depend on stimulation.

"As a society we tend to ignore the value of mental activity," Gilbert said. "We need to recognize the importance of challenging our minds as a vital component of health, of mental health."

How easily mental function can be lost was shown in the landmark Seattle Longitudinal Study. Started in 1956 by K. Warner Schaie, who is now director of the Gerontology Center at Pennsylvania State University, the study examines what happens to intellectual abilities as people age. More than 5,000 people aged twenty to more than ninety are involved in the study.

Intellectual decline varies widely, Schaie found, depending on whether people let their minds loaf or keep them busy. One out of four eighty-year-olds, for example, are as bright as they've always been. "There are very few toddling, senile millionaires," Schaie said. "It takes education and resources to make and keep that kind of money."

Couch potatoes, on the other hand, are the quickest to slip into intellectual limbo, he said. The danger starts when people retire, decide to take things easy, and say they don't have to keep up with the world anymore.

"A vicious cycle sets in," Schaie said. "If you don't do an activity anymore, you begin losing the skills to do it. Then [you] are even less likely to engage in those activities."

Bridge players do very well on mental tests; bingo players don't. Crossword puzzle workers do better on verbal skills, and jigsaw puzzle players tend to maintain their spatial skills. There are many ways to exercise the brain, but you've got to do something, Schaie said. "It's the inactive people who tend to show the most decline. The people who are almost too busy to be studied are the ones who do very well."

Seven factors stood out among those people who hung onto their intellectual prowess as they aged:

- A high standard of living marked by above-average education and income.
- A lack of chronic diseases.
- Active engagement in reading, travel, cultural events, education, clubs, and professional associations.
- A willingness to change.
- Marriage to a smart spouse.
- An ability to quickly grasp new ideas.
- Satisfaction with accomplishments.

Although some mental abilities begin to decline after the age of sixty, others, such as verbal and numeric abilities, actually increase. Schaie found that only 20 percent of adults experienced some mental decline from sixty to sixty-seven, 36 percent declined between sixty-seven and seventy-four years, and 60 percent declined from seventy-four to eighty-one.

But if mental function is easy to lose through inactivity, it is just as easily reclaimed through mental retraining. Schaie and his wife, Sherry L. Willis, found that teaching oldsters new skills increased their brain power and improved their memory, intellectual gains that lasted for years.

Their study involved 229 people over the age of sixty-five. Half had declined over the last 14 years in two important areas—inductive reasoning, or problem-solving, and spatial orientation, or knowing how to read a map or assemble objects. The other half had remained stable in these two skills. After five hours of training with geometric shapes, 50 percent of the people who had remained the same over the fourteen-year period significantly increased their reasoning and spatial abilities.

Of those people who had declined, most improved, and an incredible 40 percent of them recovered the intellectual capacities they'd had fourteen years earlier. Furthermore, when they were retested seven years after the training program, people with the biggest initial gains were still ahead of their previous levels.

The findings suggest three ways to help the brains of older people. Training can boost intellectual power, it can maintain mental function, and it can reverse memory decline and the loss of other intellectual abilities. "The results of the cognitive training studies suggest that the decline in mental performance in many community-dwelling older people is probably due to disuse and is consequently reversible, at least in part, for many persons," said Schaie.

Intellectual abilities increase with training, but what is going on inside the brain to make that happen?

Scientists know that stimulating environments greatly increase the overall number of connections in animal brains, thereby increasing brain power. Could certain exercises build up connections in specific parts of the brain to strengthen lazy or weak relay switches that keep the brain from using its full power?

To find out, Rockefeller University neuroscientist Hiroshi Asanuma trained monkeys to catch a ball. The ball was a round pellet of food that the monkey could eat if it was caught, but would disappear under the floor if missed.

The task involved two parts of the brain. One was in the sensory cortex, where information coming in through the eyes tracked the ball, and the other was the motor cortex, from which commands were sent to the arms to catch the ball.

Using sophisticated new technology, Asanuma found that as the catching task was mastered, communication lines between the sensory and motor cortex in the brains of the monkeys became stronger. More remarkable, he showed that there was a 25 percent increase in the number of connections along the sensory-motor communications highway.

No other part of the brain had an increase in connections, meaning that Asanuma was seeing for the first time the biological changes that occur when the brain learns a specific task.

"This had been expected for nearly a century but it had been difficult to prove," he said. "There's no other way of understanding learning: There must be specific changes in the brain."

Science is awakening to the fact that the brain reorganizes itself during learning, said Michael Merzenich, a pioneering neuroscientist at the University of California at San Francisco. "It's something that people don't realize. They don't think about the power that they have within themselves to change their brains.

The problem is that the brain can organize itself in good ways, like learning to play a violin or mastering calculus, or in bad ways. When the brain wires itself in negative ways the result is learning disabilities, obsessive-compulsiveness, or other behavioral problems.

When someone can't understand language, or can't read or do math, or has some behavioral problem, people tend to think it is caused by a defect in the brain. "It's not a brain defect or limitation at all," said Merzenich. "These kinds of problems really represent a different learning pathway that the brain has taken. Learning disabilities are usually learned."

Understanding that the brain can organize itself to do undesirable things is a profound insight, akin to the discovery that germs cause infection. Now neuroscientists can better understand how behavioral problems are caused, how they can be prevented, and how they can be corrected.

Merzenich, one of the first scientists to show that adult animal brains undergo massive physical changes with general learning, hopes to use Asanuma's discovery to change specific behaviors by precisely rewiring certain parts of the brain.

He already has had success with musicians suffering from focal dystonia, a condition in which the brain organizes itself to double up on muscle movements. A pianist, for instance, who continues to practice to improve his performance may suddenly find that when he moves one finger, the adjacent finger moves with it.

The two fingers can't be moved separately. The reason for this, Merzenich believes, is that in attempting to increase finger speed the brain reaches a point where the signals coming in from two fingers blend into one signal.

Focal dystonia is an example of the way the self-organizing brain can fall into a trap in which it continually reinforces inappropriate actions. Negative thoughts also become traps, recycling endlessly and producing obsessions, phobias, and other behavioral problems.

To separate the movements of the two fingers, the brain has to be retrained. Feedback devices—usually electronic instruments that measure different body responses such as muscle movement, blood flow, and temperature—are used to help the brain sense different movements in the two fingers.

"The musicians have to accomplish by dint of will and learning how to separate finger movements," he continued. "It's crucial that they know that they're actually effecting independent contractions of the finger muscles."

Behavioral therapists use the same kind of strategy to treat people with obsessions, compulsions, or other behavioral problems, even depression caused by negative thinking processes. "When you get over these things you get over them on the basis of engaging the learning process," Merzenich explained. "It involves reorganization of the brain to replace inappropriate thinking traps with more effective circuits."

A similar approach is being used to help learning-disabled children, thanks to a seminal discovery by Paula Tallal of Rutgers that the brains of these youngsters learned to process speech in slow motion. As a result, words in normal speech flow together and their meaning is lost. Tallal found, however, that when speech was slowed down, learning-disabled children could understand it.

Merzenich showed it was relatively easy to train animal brains to slow down and then speed up. First he taught animal brains to organize themselves in ways that processed sounds slowly. Then he showed that brain processing could be speeded up with new learning.

"That indicated to us that we could actually create the signature condition of a learning-disabled child and then reverse it," he said.

Using these discoveries, the two researchers and their colleagues developed a fast and simple therapy that uses computer games to dramatically improve the speech skills of children with dyslexia and other language-based learning disabilities.

"We were really shocked at the improvement in these children over a relatively short course of training," said Merzenich. In only twenty sessions that lasted twenty minutes each, children ranging in age from five to ten years were able to advance more than two years in their speech skills.

The researchers found that language comprehension improved to normal, near normal, or above normal in children who had been two to three years behind their peers in speech skills. "The children thought they were playing a game," Merzenich said. "They didn't realize that they were rewiring their brains."

The research, which was supported by a $2.3 million grant from the Charles A. Dana Foundation, has major implications for helping children with speech and reading problems and attention deficit disorder. Language-based learning problems are estimated to affect about 10 percent of school-aged children and cost the United States about $7.5 billion annually. They are a major reason why many students drop out of school.

The findings are especially encouraging for children who have difficulty learning to talk (developmental dysphasia) or have subsequent reading problems (developmental dyslexia). These disorders are notoriously difficult to treat.

"We believe this therapy would be more effective and certainly far cheaper than anything now available to treat these children," Merzenich said.

The new therapy may also help people who have suffered strokes or other types of brain damage that have impaired their ability to talk.

In a typical case, a six-year-old child whose understanding of language was like that of a three-year-old, talked like a five-year-old after the brief computer therapy.

The children are happier and more socially active. "One father told me, 'You gave me a son I can talk to,'" Tallal said. A mother reported that her daughter went to a party and for the first time was participating in the middle of the group rather than on the outside.

Girls with language disorders tend to be shy and retiring, while boys tend to be hyperactive, inattentive, and difficult to manage.

Tallal found that the reason many children have speech and reading problems is that their brains cannot process consonant sounds, which are spoken at a much faster rate than vowels. As a result, they are unable to distinguish between sounds such as "da" and "ba."

Normal children take only ten milliseconds to distinguish between two consonant sounds. But speech-impaired children require hundreds of milliseconds to recognize the difference between these different parts of speech.

It's like taking a beginning course in Spanish and then going to Spain and expecting to speak the language. For children with language disabilities, sounds run together. They don't make sense.

"Children who have trouble understanding what other people are saying to them have difficulties in developing the tools that really make us human, the ability to talk and communicate and carry on conversations," Tallal said.

Language impairment inhibits the development of reading skills because understanding the sounds of words is necessary in order for the visual words to be matched with their sounds.

Tallal's discovery—that when consonant sounds are slowed down on a recording, the children can distinguish them and learn to understand speech—paved the way for the computer therapy program.

Under the direction of William Jenkins, a UCSF professor of otolaryngology, the research team devised computer games such as "Old MacDonald's Flying Farm and Phonic Match," that the children could play on interactive multimedia CD-ROM equipment. The computer program slowed down the sound of consonants in speech and made them louder so that they could easily be distinguished by the children.

The computer finds a rate at which a child is 80 percent accurate at matching sounds to consonants or words. The computer scores every trial, and when a child gets two to three correct matches in a row, it speeds up the sounds slightly.

In the farm game, children "chase" a flying cow around the barnyard. As they do so, the computer produces a series of sounds that change unexpectedly. The child releases a button to indicate the change has been heard.

For example, a child might hear "pack, pack, pack, pat." At the sound of the last word, the child reacts.

"If he is quick enough, he gets a point and the cow runs into a barn," said Steve Miller, a researcher at Rutgers University. "If it's a miss, the cow keeps flying around."

As the children become more efficient at recognizing individual sounds, the computer continually speeds up the sounds, providing rewards every time the children notch a higher speed.

Within a short period of time, the children are able to comprehend speech at near normal or above normal levels, indicating that their brains are being reorganized as a result of experience and training.

"That's quite amazing," said Tallal. "It's very hopeful because it suggests that for whatever reason these children are impaired—and end up with language and reading problems—it's fixable."

The improvement appears to be permanent, because the brain of a learning-impaired child is hungry for this kind of information. It is brain food the child has not tasted before.

"Now that they are exercising those parts of the brain that process faster information, they're getting that information in the real world all the time now, which helps them to constantly strengthen their language capabilities," Tallal said.

Learning-disabled children are often misdiagnosed as having an attention deficit disorder. The reason these children do not pay attention is because they don't understand what the teacher is saying.

Tallal and her colleagues are developing screening tests to identify infants who may be susceptible to language disorders, so that early treatment might head off disabilities.

"Our hypothesis is that these children may never have these problems if we can provide them with the kind of speech that they can process and strengthen those parts of their brains that process fast combinations of sounds," she said.

The actual underlying cause of speech impairment is unknown. But researcheres believe that many things may contribute to the problem. Among them are ear infections that block normal hearing during a critical period of brain development, or a child simply learning the wrong way to hear sounds.

"There's really nothing wrong with the learning machinery in these kids," said Merzenich. "It's just that they learn in a way that is counterproductive for dealing with fast speech.

"This represents the beginning of a revolution—a clearer understanding that learning and self-organization of the brain have an impact not just for learning-disabled children, but for everybody."

16 Reversing Stroke and Spinal Cord Damage

Scientists are on the brink of doing the unthinkable—replenishing the brains of people who have suffered strokes or head injuries to make them whole again. And as if that is not astonishing enough, they think they may be able to reverse paralysis.

The door is at last open to lifting the terrifying sentence these disorders still decree—loss of physical function, cognitive skills, memory, and personality—which costs the nation $65 billion annually.

Until recently there was virtually nothing doctors could do for the 500,000 Americans who have strokes each year, the 500,000 to 750,000 who experience severe head injury, or the 10,000 people who are paralyzed after spinal cord damage.

But that is about to change. Researchers now think it may be possible to replace destroyed brain cells with new ones to give victims of stroke and brain injury a chance to relearn how to control their body, form new thinking processes, and regain emotions.

And after demolishing the long-standing myth that brain cells can't regenerate or proliferate, scientists are developing ways to stimulate cells to do just that.

Although stroke, head injury, and paralysis are three of the most devastating things that can happen to anyone, scientists have recently learned that the damage they cause is not preordained: it takes place over minutes, hours, and days, giving them a precious opportunity to develop treatments to halt much of the damage.

Most of the new remedies are not yet available, but an explosion of research in the last five to ten years has convinced scientists that some of them will work.

Buoyed by fabulous results in preventing permanent damage from stroke and other injuries to the central nervous system in rats and other animals, researchers around the world have launched scores of trials in humans.

But many promising new therapies are sitting on the shelf because of a lack of money and other resources necessary to conduct large, lengthy, and expensive studies to conclusively show that a new drug or treatment really works in people.

"We could treat a rat who came into our hospital with a stroke very successfully today," said Washington University neurologist Dr. Dennis Choi. "Hopefully we're not very far away, in terms of scientific proof, from the day we can use the same kinds of strategies that are saving rats to save people. Today this hospital will have a couple of patients admitted with acute stroke and we won't be able to use those treatments yet because they haven't been proven safe and effective in humans."

The requirement for safety and efficacy can be frustrating, especially for badly needed treatments that are very promising. But such caution is necessary.

No one knows that better than Ray Wickson, a paraplegic who is president of the Canadian Spinal Research Organization. The organization is trying to raise funds for a major study to test a pill for paralysis—4 amino pyridine (4 AP)—which has shown promising results in preliminary human trials.

About half of the small number of people in the study, who had been paralyzed for four to fifteen years, regained some sensation and muscle function when they were given intravenous infusions of 4 AP. (In the new study the drug will be taken in pill form.) The benefit lasted for days in some people.

Although the effect gradually faded, the doctors and patients remained amazed.

"Spinal-cord-injured people give it up and say 'this is the way it's going to be for the rest of my life,'" Wickson said. "When something comes along and changes the biology of the injury and then you can move something, or you can feel something, it takes time to comprehend what's happened. It's kind of a stunned reaction."

For Purdue biophysicist and embryologist Richard B. Borgens, who, along with Andrew Blight of the University of North Carolina, showed that 4 AP could restore some degree of function in dogs that were paralyzed in accidents, the early Canadian results "made our mouths drop."

"4 AP can reverse deficits that have lasted for years," Borgens said. "It reverses them in minutes. The significance of that is that it can reverse not only paralysis but a dogma of hopelessness."

A similar story is unfolding in traumatic head injury. Preliminary results from early studies have shown that using cooling blankets to lower temperatures four to five degrees improves recovery by 15 to 16 percent, said Dr. Guy Clifton, chief of neurosurgery at the University of Texas Health Science Center in Houston.

But as promising as the small studies are, they are not convincing enough to mandate all emergency rooms to cool head-injury patients who are comatose. A major study is needed to prove that it really works and that it should become standard care.

After years of wondering about whether cooling a patient could dramatically reduce the damage from head trauma, the National Institutes of Health has decided to fund a major nine-center study.

"Brain injury research was considered a squashed bug ten years ago," said Clifton, leader of the NIH study. "People who worked in this area, like myself, were considered to be involved in a futile undertaking. Now you can just taste the potential. We're going to see stuff that nobody would have believed ten years ago."

Scientists are finding that treatments that work in one type of injury—stroke, head trauma, or spinal damage—are likely to work in the others. All of these disorders share many of the same mechanisms of cell destruction, which come in two phases, primary and secondary injury.

In the primary, or initial, injury, blood flow to a part of the brain is blocked by a clot that plugs an artery or by a physical blow. Brain cells, or neurons, are either damaged or die right away because they are deprived of nourishing blood.

This initial destruction then triggers a chemical attack against tissue that was not damaged in the primary injury. The second phase of injury invokes a process called excitotoxicity and it affects nearby healthy cells, often killing more brain tissue than the initial injury.

Like someone yelling "fire" in a crowded theater, damaged and dying cells scream out a slew of chemicals. These chemicals, which normally help brain cells talk to each other, become dangerously toxic in excessive amounts. They literally cause healthy cells to become overexcited to the point of death, when they too spew out their death-throe chemicals.

Interestingly, scientists believe that excitotoxicity is a genetically programmed suicide mechanism devised by nature to kill unneeded or unhealthy cells. Such cell death occurs during fetal development, for instance, to get rid of billions of overproduced brain cells and the webbing between fingers.

It is this same excitotoxic response that is rapidly triggered in stroke, head trauma, or spinal injury to produce the destructive secondary injury. Evidence also indicates that the excitotoxic reaction can occur over a longer period of time, causing a slow form of suicide that may be the final pathway for cellular death in Alzheimer's, Parkinson's, and other degenerative neurological disorders.

The suicide reaction—its scientific name is apoptosis—begins when a damaged or dying neuron releases massive amounts of a neurotransmitter called glutamate. Glutamate is normally one of the most important chemical messengers in the brain.

But when too much glutamate is present, the NMDA receptors ("doors" on cell surfaces) are jammed open. Sodium floods in, causing the cell to swell. Calcium rushes in and smashes at the cell's genetic controls, producing enzymes that eat away the cell's internal support structure and destructive molecules, called free radicals, that chew away its membrane wall.

"It would be like going into the cabin of a 747 jetliner with a sledgehammer and starting to hit left and right," Washington University's Dennis Choi said. "Everything just starts going haywire."

The discovery of the key steps in the suicide cascade of secondary injury is leading to the development of drugs to block them. Experiments in animals show that by blocking the secondary injury, much of the damage that normally occurs from a stroke, head trauma, or spinal injury can be prevented.

Two drugs that block the NMDA receptor, thereby preventing glutamate from entering a cell and starting the suicide reaction, have been found to be 40 to 70 percent effective in preventing brain damage from stroke in animals.

The drugs, Cambridge Neuroscience's cerestat and Ciba-Geigy's selfotel, are in clinical trials with people who have had strokes or severe head injuries.

Another drug has been designed to prevent damaged neurons from releasing glutamate and still others are intended to block sodium (lubeluzole) and calcium (nimodapine) from flooding into healthy cells.

Upjohn's Tirilazad, which neutralizes cell-damaging free radicals, has been approved for use in male stroke victims in five European countries after successful clinical trials there. The drug was only marginally effective in women, perhaps because females eliminate the drug too swiftly from their bodies. Trials using higher doses for both men and women are under way in the United States.

Scientists are also trying to disarm the enzymes that chomp away at the inside of neurons. Boston-based Alkermes has developed a drug that dramatically blocks enzymatic destruction of cells in animal models of brain injury.

As good as some of these new drugs are, they are even better when used in combination, said Dr. James C. Grotta, director of the stroke program at the University of Texas Health Science Center in Houston.

Animal studies have shown that combining two or more of the promising drugs that block secondary injury, or mixing them with a clot-busting medication called TPA or with whole body cooling produces better results than when each is used alone. Cooling slows cellular function, thereby preventing the production of dangerous levels of glutamate, which starts the secondary injury response.

"We are on the threshold of a major improvement in stroke treatment," Grotta said. "Everyone believes strokes can be treated with some of what we know already from animal experiments. It's simply a matter of translating that into reality through effective clinical trials."

How quickly that threshold is being crossed was demonstrated recently by the startling results of a study showing that TPA can dramatically reduce brain damage from stroke, the main cause of adult disability and the third leading cause of death in the United States.

The study, which was sponsored by the National Institute of Neurological Disorders and Stroke, is a major breakthrough, providing hope where there was none.

Receiving TPA within three hours of a stroke increased by 55 percent the chances that a patient would recover with little or no brain damage. The window of opportunity is important because giving TPA after three hours could cause dangerous bleeding in the brain. The study, which involved more than six hundred patients, was conducted at nine major medical centers.

Just as emergency treatment of people with heart attacks and spinal injuries was greatly speeded up once effective therapies were developed, people who suffer strokes or brain trauma will have to be taken to medical centers as quickly as possible once these new treatments become part of standard care.

Many people don't even recognize that they are having a stroke because they don't know the symptoms—major numbness or tingling that comes on suddenly on one side of the body or the other, loss of vision in one eye or the other, clumsiness, weakness, and difficulty speaking or understanding speech.

About 85 percent of strokes are caused by clots that form in the brain or that drift there from the heart or other sites. Clots prevent oxygen-rich blood from reaching brain tissue, which quickly dies. Breaking up clots as fast as possible is important because the brain, although it makes up only 2 percent of the body's mass, consumes 20 percent of its oxygen. Fifteen percent of strokes are caused by ruptured blood vessels.

TPA may be effective in 50 percent or more of the patients whose strokes are caused by clots. Giving TPA to patients who have suffered bleeding strokes from torn vessles would be dangerous because it could increase the risk of more bleeding. Physicians use CAT scans, which provide images of the brain, to determine which type of stroke has occurred.

Patients with head injury are being helped with a potent antioxidant drug called PEG-SOD. Early results show that PEG-SOD improves favorable outcome of head-injury patients by 18 percent. More of them are able to take care of themselves and even go back to work.

While some scientists are developing ways to reduce the damage from brain injuries, including stroke, others are trying to prevent them. The simplest and most startling discovery is that eating plenty of fruits and vegetables can dramatically reduce the risk of stroke.

Fruits and vegetables contain important vitamins and other nutrients that help keep blood vessels in the brain healthy by neutralizing molecules called free radicals. Free radicals can damage tissue if they are not disarmed by vitamins that act like antioxidants. A person who eats more than eight servings of fruits and vegetables a day reduces his or her risk of suffering a stroke by 60 percent compared to a person who eats less than two servings.

Among the most protective nutrients found in fruits and vegetables are folate and vitamins B_6 and B_{12}. Jacob Selhub, a Tufts University nutritional biochemist, found that low levels of these three nutrients produce slightly elevated levels of an amino acid called homocysteine.

Homocysteine in small amounts is crucial for proper metabolism. But a slightly elevated level of homocysteine, for some still unknown reason, appears to be a major new risk factor for stroke, increasing by three to five times the likelihood of having a stroke. Elevated homocysteine also increases the risk of heart disease by three and a half times. Discovery of the homocysteine risk helps explain the mystery of why more than half of the people who suffer heart attacks have normal cholesterol levels.

Other prevention strategies that have been proven to be effective in clinical trials include:

- Surgically removing fatty blockages in the two main blood vessels that feed blood to the brain, which reduces the risk of stroke by 55 percent.

- Treating an erratic heartbeat known as atrial fibrillation, which generates free-floating blood clots and causes 70,000 strokes a year. The risk from this type of stroke can be reduced by 50 to 80 percent with two blood thinners—aspirin and warfarin—which prevent clot formation.

- Replacing estrogen that sharply declines in women after menopause, which can cut the risk of stroke in women by 50 percent.

The turning point in brain-injury research came in 1990 when a team of researchers led by Dr. Wise Young of the New York University Medical Center reported beyond a shadow of a doubt that a drug called methylprednisolone could significantly reduce paralysis from spinal cord injury.

The study sent shock waves through the medical world, shattering the firmly held belief that nothing could be done to save injured neurons.

When given to patients within six hours of a spinal injury, methylprednisolone increased recovery of function by 20 percent. The drug acts as both a powerful antioxidant and an anti-inflammatory agent.

"The methylprednisolone study was a tremendous boost," said Dr. Michael Walker, director of the stroke and trauma division of NIH's Neurology Institute. "It suddenly said there is something we can do about this spinal cord injury that we thought was untreatable.

"From a pharmacological point of view, there are really very few drugs like methylprednisolone, where its use for just twenty-four hours changes the patient's life for the rest of his or her life. It has that profound an effect. It's really quite extraordinary."

The care of people with spinal cord injuries has changed quickly. Before 1990, when it was thought there was absolutely nothing that could be done for these patients, medical personnel did not hurry to treat them. These patients would often lie in emergency rooms for hours until a neurosurgeon could be found to look at an X-ray.

Now, people with spinal cord injuries are treated as rapidly as are people with heart attacks. Paramedics are equipped with spine boards to stabilize patients and prevent further injury. They are given methylprednisolone immediately.

Superman star Christopher Reeve, who broke his neck when he was thrown from his horse, got the drug after his injury and so did

New York Jets lineman Dennis Byrd. Byrd suffered a football injury to his spine in 1992 that, without methylprednisolone, probably would have left him about 50 percent paralyzed. Today he is jogging.

The use of methylprednisolone has dramatically improved the outcome of spinal cord injuries. Prior to 1990, 67 percent of patients with spinal trauma had no sensation or movement below the point of injury. Today, that figure has been reduced to 37 percent; thousands of patients are recovering more function and are able to take care of themselves.

"Our methylprednisolone study cost the U.S. government $6 million over a five-year period," Young said. "For that $6 million we are probably saving the U.S. government a minimum of $600 million a year in money that doesn't have to be spent on caring for people with complete paralysis.

"A critical mass is developing. Within the next five to ten years we will have effective therapies for chronic spinal cord injury."

Reversing it?

"Yes. Absolutely."

But getting those therapies to market will be tough. Drug companies have little interest in medicines for spinal cord injuries because, with 10,000 victims a year, there is little money to be made. They are more interested in drugs for stroke and head injuries, which affect hundreds of thousands of people annually. And dwindling NIH resources make it unlikely that the government can fund clinical studies of many of the therapies now sitting in laboratories.

So Young and some of the world's top neuroscientists are taking matters into their own hands. They have set up a company, Acorda, to raise funds for clinical trials.

At the top of their list are drugs to revive cells that sulk after injury. Much of the paralysis from spinal injury is not caused by nerves that die, but by nerves that can't send electrical signals anymore because their insulating myelin sheaths have been damaged.

Neurons communicate by extending biological telephone lines called axons to other neurons. An axonal line, or fiber, that connects one nerve to many others becomes short-circuited when its myelin insulation is torn off, like an electric cord that malfunctions when its copper wire is exposed. This process, called demyelination, is the same problem that causes loss of function in people with multiple sclerosis.

Acorda plans to test a number of new drugs that may enable damaged neurons to regrow their myelin covering. In fact, the paralysis-reversing drug 4 AP appears to work by wrapping a chemical tape around exposed axons to prevent short circuits and allow electrical signals to complete their course. In addition to its early encouraging results in paralyzed patients, 4 AP appears promising for treating multiple sclerosis.

Paralysis can also result when axons break their connections to other cells. A number of compounds are able to stimulate the regrowth of axons in laboratory studies, and Acorda is interested in pushing these into human trials.

The ability to cause neurons to regenerate their axons was thought to be an unattainable goal. But Martin Schwab, chairman of the department of morphology at the University of Zurich's Brain Research Institute, suddenly brought that goal within reach.

Schwab discovered that the reason neurons don't regenerate is because they are forbidden to do so by an inhibitory factor. When he removed the inhibitory chemical with an antibody that neutralized it, and then doused neurons with a growth factor called NT3, they regrew their axonal fibers.

In a spectacular series of experiments, Schwab showed for the first time that his technique could regrow axons the whole length of an adult rat's spinal cord. The axons grew from the site of injury and reconnected with their target cells, restoring some degree of function.

"Schwab really opened the door to regeneration of the spinal cord," Young said. "His work turned the whole field on its ear. Nobody wanted to work in this area because they thought it was impossible. Now, because of Schwab, hundreds of laboratories are working on this subject."

The findings shed new light on why nature said "no" to regeneration in the first place. The brain is equipped with many different kinds of growth factors to help neurons get wired during early development into patterns that can then execute movement, learning, and memory. Once those networks are established, inhibitory factors are manufactured to make sure they stay in place.

Schwab describes the process as similar to an artist creating a mosaic by setting individual tiles into a pattern. After the picture takes shape, he pours cement around the tiles to lock them in position.

"We know these inhibitory factors are also present in the brain," Schwab said. "When we remove them we see a regenerative response from the damaged nerve tracts."

Other strategies to save neurons are being tried and are producing remarkable results in the laboratory. Neuroscientist Lloyd Guth of the College of William and Mary is able to get rats to walk after what otherwise would be a paralyzing spinal injury.

He treats them with three chemicals. Each chemical alone has no effect. But when combined they cause a synergistic reaction that enables the animals to begin walking after ten days. The drugs are lipopolysaccharide (LPS), which stimulates the production of brain-cell growth factors; indomethacin, which cools down the anti-inflammatory response produced by the injury; and pregnenolone, a basic hormone that gives rise to many steroids that are used for repair.

Dr. Jewell Osterholm, a neurosurgeon at Thomas Jefferson University, has the most direct approach for stroke and head trauma. He flushes out the brains of animals within three hours after experimentally induced strokes, preventing as much as 95 percent of the usual damage.

His flushing fluid is an emulsion of tiny fluorocarbon globules that contain oxygen and other nutrients that are important for brain metabolism. The globules are one-hundredth the size of red blood cells and permeate the entire brain, flushing out glutamate, sodium, and calcium, the three chemicals responsible for the secondary injury to brain tissue.

The emulsion is pumped into the ventricle cavities in the center of the brain through a tube that is inserted into a tiny hole drilled into the skull. Another tube is inserted into the spinal cord to drain the emulsion.

Repairing damaged neurons with gene therapy is the goal of Dr. Ronald Hayes, director of the neurosurgery research laboratory at the University of Texas Health Science Center in Houston. He uses minuscule fat globules packed with genes that regulate the production of such growth factors as NGF and BDNF, which stimulate cell repair.

The tiny globules are absorbed by injured brain cells, and the genes inside manufacture the healing growth factors that do such things as rebuild myelin sheaths and stimulate the regrowth of axons.

"We've shown in our tissue culture system that you can put these genes into neurons that are injured but surviving and then they repair themselves by producing critical structural proteins that were destroyed by the injury," Hayes said.

Scientists are hopeful that fetal transplants will also work to repair damage from stroke, head trauma, and spinal cord injury. Animal research shows that fetal transplants not only can replace damaged brain tissue, but they can restore the ability to learn and remember.

Scientists are even learning how to make new brain cells. Some of the most exciting research is being done by researchers who are coaxing dormant stem cells back into activity. Stem cells are the progenitor cells that give birth to all the other different brain cells during fetal development. It was once thought that they vanished when their job was done. Now it is apparent that, for the most part, they become dormant and can occasionally become active again to create new neurons.

Neuroscientist Elizabeth Gould of Rockefeller University is able to stimulate quiescent stem cells to begin multiplying again by temporarily turning back the clock to the time when the brain was being built.

During early development brain cells are born in vast numbers, but they are not yet communicating with one another in any major way. Stem cells seem to want to make more brain cells when the brain is quiet. Once stem cells hear neurons chattering among themselves, they think their job is done and stop making new neurons.

Gould mimics early development by quieting brain cells with a drug called MK801. The drug turns off the key "on" switch—the NMDA receptor—that tells cells to fire to talk to other cells. As brain-cell talk ceases, stem cells perk up and start making more neurons.

"In time it results in a net increase in the number of neurons," Gould said. "These cells aren't just being born and then dying away. They're actually surviving and being incorporated into functional circuits."

The new cells are being born in the hippocampus, which is intimately involved in memory formation. Perhaps the birth of new cells occurs when a new capacity for learning is needed.

When large numbers of brain cells are needed to replace those destroyed by stroke or brain trauma, scientists now think they know where to get them. In the near future there may be "supermarkets" for neurons, based on the discovery of ways to make fetal brain stem cells multiply in almost unlimited amounts.

Researchers are also figuring out how to get fetal stem cells to differentiate into the specialized cells that are located in different parts of the brain. Growth factors such as "hedgehog," FGF, NT3, and BDNF are being used to turn undifferentiated neurons into specialized cells. For example, they can convert stem cells into neurons that make dopamine, the neurotransmitter that enables muscles to work, or cells that process memory.

"We have the stem cell jumping through hoops," said the Neurology Institute's Ronald McKay. "We have demonstrated that you can expand the cells and then differentiate them to particular neuronal fates. They can give rise to all the cells of the brain."

In a groundbreaking experiment, McKay, Anders Bjorkland from Sweden's University of Lund, and their colleagues showed that they could convert stem cells into dopamine-producing neurons.

Implanted where they normally reside in the brain the neurons grew axonal connections to the striatum, the part of the brain to which they would normally deliver dopamine to control muscle movement.

"We're making tremendous leaps and bounds from that dark period not long ago when we thought nothing was possible," said New York University's Young. "The key to all this is the fact that the vast majority of scientists now believe that the central nervous system can regenerate. They are now convinced that it's only a matter of time before we discover how to make it grow."

Epilogue

The amazing discovery of the brain's plasticity—its ability to physically rewire itself to become smarter—makes mental stimulation, in the long run, more essential to the body than food.

That the brain thrives with good nourishment is a concept that has profound significance for individual achievement and for the way parents raise their children. The brain's food is education. Just as the food we eat gives our immune systems the strength to fight off life-threatening infectious germs, education protects us against bad choices. In effect, education acts like a vaccine that boosts our mental powers, making us more resistant to illness and premature aging.

Education provides such strong immunity, in fact, that people who acquire more of it are living longer than ever before while those who don't have it are falling farther behind. It is the secret to a healthier, longer life.

Scientists have long known that income, occupation, and education are the most important predictors of people's health and how long they will live. But they had no way of telling which had the biggest impact.

Income had been the front-runner. Who can argue with the late singer Sophie Tucker's observation: "I have been rich and I have been poor. Rich is better." And a good job brings self-esteem and other rewards.

But it is education that is emerging as the most critical predictor of longevity and good health. It's what you don't know that can hurt you.

That comes as good news to public health experts who are desperately looking for ways to head off an approaching tidal wave of sickness and mental dysfunction that is facing a rapidly aging U.S. population.

For people who don't grab at the opportunity for education, the news is grim. They are on the wrong end of a widening gap between people who build more brain power and those who ignore it, and they are more likely to die younger.

Despite an overall decline in death rates in the United States since 1960, poorly educated low-income white males die at rates that are three to seven times higher than white men with better education or higher income, Dr. Gregory Pappas of the National Center for Health Statistics found in a 1993 study. The findings were reported in the *New England Journal of Medicine.*

This disparity increased between 1960 and 1986 as better-educated people were quicker to respond to the growing scientific evidence of the dangers of smoking, high-fat diets, and physical inactivity, and to adopt a healthier lifestyle.

The payoff has been remarkable. There are one-third fewer cigarette smokers, and the consumption of cholesterol and salt has declined dramatically. These lifestyle changes are largely responsible for a 53 percent decline in heart attack deaths since 1963.

Those who benefited the most were the better educated. During the twenty-six-year period in the Pappas study, the age-adjusted death rate for white male college graduates twenty-five to sixty-four years of age declined by 50 percent from 5.7 deaths per 1,000 men in 1960 to 2.8 deaths per 1,000 in 1986. Among white men who did not finish high school, death rates dropped by only 15 percent, from 9 to 7.6 per 1,000.

"If there were some infectious disease in the population that caused such a differential in mortality, it would be a national disaster," said Dr. Jack M. Guralnik of the National Institute on Aging.

"If a large proportion of the population had something that increased its mortality rate by these amounts, we'd say 'My God, this is horrendous,'" he said. "But nobody pays much attention to this. Certainly the medical establishment doesn't."

Unlike money, which is hard to come by, and professional status, which is difficult to achieve, education can be acquired by anyone.

Its power lies in the mind—the ability to convert words, ideas, and new information into problem-solving actions.

"In all of the studies looking at lower mortality and lower morbidity rates, people who do much better are not an elite group," Guralnik said. "They're high school graduates. Just getting out of high school puts people in a category where it looks like their risks of morbidity and mortality are substantially lower." People with twelve or more years of education can look forward to nearly four more years of active life than those who are less educated.

The earlier education is acquired, the more impact it has against sickness and early death. But throughout life education acts like a continuing series of booster shots. Education works in two fundamental ways:

- Biologically, by laying down significantly more connections between brain cells that accompany learning. Memory, as a result, is increased and the additional connections also provide a buffer against the destructive forces of Alzheimer's disease.

- Behaviorally, by promoting positive values and attitudes about health, higher self-esteem, effective coping skills, access to preventive health services, and association with people who have similar views. At the same time, education reduces risky behaviors such as smoking.

The lack of education, though not as visible as an infectious disease, is far deadlier in the industrialized world. The death rate disparity that Pappas found between white men, women, and African-Americans occurred at a time when educational levels increased dramatically in some groups, while hardly budging in others.

The growing gap between the educated and less educated is reflected in literacy rates. In 1960 the U.S. government pronounced someone literate if he or she had an eighth-grade education.

That standard of literacy has become hopelessly out of date. Today the National Center for Education Statistics defines literacy as ". . . using printed and written information to function in society, to achieve one's goals, and to develop one's knowledge and potential."

Clearly, being literate means more than just being able to read at an eighth-grade level. Information has to be sucked in and used. And while the number of formal years of education serves as an average predictor of how well a person will do, it really depends on how he uses his brain all of his life.

Just as some people fail to get vaccinated against common childhood infections, others fail to take advantage of the immunizing effects of education. Half of the high school students in Chicago and some other large cities, for instance, fail to graduate.

The toll this takes on the brain is staggering. Children born to mothers who have less than twelve years of education have a fourfold increased risk of mental retardation, said Dr. Marshalyn Yeargin-Allsop, a medical epidemiologist at the Centers for Disease Control and Prevention's (CDC) Division of Birth Defects and Developmental Disabilities.

"This is regardless of race," she said. "White children and black children had the same fourfold risk if their mothers didn't complete high school."

A CDC study of more than 1,000 children showed that mild retardation, defined as having an IQ between 50 and 70, occurs at the rate of nearly 1 in 100 children. The biggest risk factor for mild retardation is the mother's low educational level, which far exceeds the risk posed by poverty.

About 22 percent of all births in this country are to mothers with less than a high school education. These women often do not know how to provide stimulation—talk, toys, physical activity—to their infants, which can lead to stunting of the brain during the crucial first three years of life, Yeargin-Allsop said. "We're wasting a lot of human potential. It's astounding to think that almost a quarter of all children are being born to mothers with less than twelve years of education and are at risk for mild mental retardation."

Mild mental retardation is generally believed to be caused by a failure to provide the brain with the kinds of experiences from its surrounding world that it needs to develop to its maximum capacity. Early educational intervention programs for children at risk have shown that they can increase IQ levels by fifteen points or more.

"At least half of the cases of mild mental retardation are preventable," Yeargin-Allsop said. "We can leapfrog over the risks if young people stay in school and get as much education as they can."

Based on these findings the CDC is planning to launch Project BEGIN, in which those children from birth to age three who are at risk for retardation will participate in intensive intervention programs. The children will be enrolled in full-day, year-round child development centers, and their families will have regular home visits by psychologists and social workers.

Having less than twelve years of education makes people more vulnerable to other risks, such as ad campaigns that increase risky behavior, said Stanford University epidemiologist Marilyn A. Winkleby. "Tobacco companies are savvy about this. They've developed all sorts of ads, as have alcohol companies. They've done their market research and target people with low education. Health professionals need to be as good about doing their market research to counter harmful lifestyles."

Some people, on the other hand, are on an information binge. The average newspaper reader, for instance, has an education level equivalent to one year of college.

The dramatic difference in mortality and morbidity between the educated and less educated occurs around the world. The disparity can be found in such countries as Sweden, Canada, and England, which have national health insurance, a benefit that presumably gives all citizens equal access to health care.

Johns Hopkins researchers were stunned to find the same thing in their own backyard. The biggest cause of blindness in Hopkins's East Baltimore low-income neighborhood is cataracts. Such blindness is easily reversed by simple surgery to remove the cataracts, yet these people did not know it was available at nearby Hopkins and that the cost would be covered by Medicare.

"They were blind," said Dr. Alfred Sommers, dean of Hopkins's School of Public Health. "We're not talking about having some difficulty reading. We're talking about people who couldn't see TV and couldn't read. A very large component has to be their educational level, because that's what changes their motivation and understanding about what's going on."

Sociologists and public health experts have traditionally looked at money, job status, and education as the main determinants of health. They have even examined ethnicity, poverty, stress, crime, and other factors generally ascribed to social class differences.

For a long time many people believed that race accounted for the higher sickness and death rates among some groups. New research tends to dispel that connection as a major concern.

"Certainly there are cultural differences, but the big, big indicator is low educational attainment, regardless of ethnicity," said Winkleby.

Whites live longer than blacks primarily because proportionately more blacks have lower education and income levels, Guralnik said. When educated blacks are compared with educated whites, there's no difference. "Race is not what's important. It's socioeconomic status that is driving this. And independent of the other aspects of socioeconomic status—income and occupation—education, in and of itself, probably does play a positive role in improving health status," he said.

Many scientists believe that education affects health because it helps people determine how to live their lives, and whether to choose to take risks or not.

Lifestyle risks account for half of the 2.2 million deaths that occur annually in the United States, according to Dr. Michael McGinnis of the U.S. Department of Public Health and Dr. William Foege of the Carter Presidential Center in Atlanta. In an eye-opening 1993 article published in the *Journal of the American Medical Association* the researchers described the deadly toll: tobacco, 400,000 annual deaths; bad diet and physical inactivity, 300,000 deaths; alcohol, 100,000; infections (mostly preventable), 90,000; toxic agents at home or in the workplace, 60,000; firearms, 35,000; unsafe sex, 30,000; motor vehicles, 25,000; and illicit use of drugs, 20,000.

"Lack of education is not only a very powerful risk factor, it's a modifiable risk factor," said Winkleby, who found that with each additional year of schooling there was a dramatic decrease in heart disease risk factors. "With education you learn how to navigate your world. You learn empowerment. You learn how to articulate your needs and to overcome potential barriers."

Those barriers can be formidable, as the first nationwide study to assess lifestyle risk factors by sex, race, ethnicity, and education recently found. The study, which involved more than 180,000 adults, was sponsored by the CDC.

Its chief finding hammered home the conclusion of smaller studies: as education levels rise, the risks of smoking, sedentary behavior, and obesity decline.

For white men, 25 percent of those with less than twelve years of schooling smoked compared to 17 percent of men with more than twelve years. Nearly three fourths of the men with less than twelve years of education had a sedentary lifestyle compared to less than half of those with more than twelve years. Thirty-two percent of men with less than twelve years of schooling were overweight compared to only 17 percent of those with more than twelve years.

A similar pattern was found for white women, blacks, Native Americans, Alaskan natives, Asians, Pacific Islanders, and Hispanics.

"When you find study after study showing this relationship between health and education, then education becomes a defining factor," said CDC epidemiologist Nora Keenan.

What is it about education that makes a difference? When David S. Strogatz of the State University of New York and the New York State Department of Public Health put the question under his epidemiological microscope he found that education allowed people to recognize risks and do something about them.

The odds of college graduates being able to state their own blood pressure and to know that 140/90 mmHg or less is a "good blood pressure" were more than three times greater than for people who had not completed high school, he showed in a study of more than 3,000 people.

College graduates were more than three times more likely to know their own cholesterol level or to know that 200 mg|dl or less is a healthy cholesterol level. The odds of college graduates being current smokers were less than half as great as for those not finishing high school.

In contrast, when poor, rural Appalachian patients were asked why they didn't take advantage of blood pressure testing, cholesterol screening, Pap smears, mammography, and other preventive measures, 51 percent said it was because they didn't know why they should. Thirty-six percent listed cost as the reason.

If early education is like a vaccine against risk factors, then giving it to poor, deprived children should help protect them as they grow older. That's what David Weikart, president of the groundbreaking High/Scope Perry Preschool Study in Ypsilanti, Michigan, set out to prove in the late sixties.

Weikart randomly divided 127 black children ages three and four into two groups. The children were born in poverty and had a high risk of failing in school. One group received intensive preschool education 2.5 hours a day for thirty weeks. The other group, which received no intervention, served as controls.

"The educational program focused on getting kids to make choices between things that might either be good or bad for them and to invent solutions to the problems they were working on," he said.

Now, twenty-seven years later, the children who were in the intervention program are doing significantly better as adults than the controls. Seventy-one percent of them finished high school compared to 54 percent of the children in the control group.

And they have less risky lifestyles. The children who were in the education program have less than one-third the risk of being arrested for drugs as the controls, and girls in the program are one-third less likely to have babies out of wedlock than their peers in the control group. Those in the program are less likely to have been on welfare and more likely to own their own homes, have good jobs, and be in a stable marriage.

"For every dollar we initially spent on educating the children in the intervention group, the public is receiving $7.16 in savings from reduced crime, reduced welfare, reduced cost of education, and more tax payment on earnings," Weikart said.

Just as Weikart set out to prove that early education can produce positive lifelong behavioral changes, David A. Snowdon, associate professor of preventive medicine at the University of Kentucky, wants to show that childhood educational experiences can biologically build a better brain.

One of the problems with assessing the impact of education on health is adjusting for things like smoking, drinking, access to health care, and other factors that can muddy results.

By studying 678 nuns, members of the School Sisters of Notre

Dame, who were born in 1916 or earlier, Snowdon was able to eliminate most of the other risk factors. From the age of twenty on, the nuns shared the same environment, ate the same food, had the same access to health care, and didn't smoke or drink.

Their only major differences were the level of education they had when they became nuns and the type of job they performed during their career, such as teaching or housekeeping.

So far his findings are right on target. The more highly educated nuns live four years longer in good mental and physical health than those who have less than a bachelor's degree. The less educated sisters have twice the death rate at every age between twenty and ninety-five.

Childhood educational skills are a key to predicting which nuns will live the longest. How well the nuns wrote and used vocabulary early on in their lives indicated the level of intellectual capacity they had achieved, which showed their willingness to use their brains.

"We find that sisters who had a limited vocabulary when they were twenty are the ones at high risk of mental and physical disabilities sixty to seventy years later," Snowdon said.

Growing evidence indicates that early mental stimulation promotes the growth of synaptic connections between brain cells. The brain has trillions of these connections, which act as telephone lines that enable cells to talk to each other for the purposes of creating memory and intellectual prowess.

"So, building a better brain, keeping your brain active, particularly starting at a very young age, might offer you some protection against brain disease and disabilities later on," Snowdon said.

The sisters think it's worth finding out. All have agreed to donate their brains for study after their death to determine if the brains of the better-educated nuns actually have more synaptic connections than the brains of those who are less educated. So far, autopsy studies of the brains of nuns who have already died point in that direction.

"Sisters with less education have smaller brains at death," Snowdon said. Why? "It might be because they didn't lay down as many synapses early on as the sisters who were more intellectually stimulated."